David Achermann

# Maintenance Strategies under Consideration of Logistic Processes

David Achermann

# Maintenance Strategies under Consideration of Logistic Processes

## Modelling, Simulation and Optimization

Südwestdeutscher Verlag für Hochschulschriften

**Impressum/Imprint (nur für Deutschland/ only for Germany)**
Bibliografische Information der Deutschen Nationalbibliothek: Die Deutsche Nationalbibliothek verzeichnet diese Publikation in der Deutschen Nationalbibliografie; detaillierte bibliografische Daten sind im Internet über http://dnb.d-nb.de abrufbar.
Alle in diesem Buch genannten Marken und Produktnamen unterliegen warenzeichen-, marken- oder patentrechtlichem Schutz bzw. sind Warenzeichen oder eingetragene Warenzeichen der jeweiligen Inhaber. Die Wiedergabe von Marken, Produktnamen, Gebrauchsnamen, Handelsnamen, Warenbezeichnungen u.s.w. in diesem Werk berechtigt auch ohne besondere Kennzeichnung nicht zu der Annahme, dass solche Namen im Sinne der Warenzeichen- und Markenschutzgesetzgebung als frei zu betrachten wären und daher von jedermann benutzt werden dürften.

Verlag: Südwestdeutscher Verlag für Hochschulschriften Aktiengesellschaft & Co. KG
Dudweiler Landstr. 99, 66123 Saarbrücken, Deutschland
Telefon +49 681 37 20 271-1, Telefax +49 681 37 20 271-0, Email: info@svh-verlag.de
Zugl.: Zürich, ETH, Diss., 2008

Herstellung in Deutschland:
Schaltungsdienst Lange o.H.G., Zehrensdorfer Str. 11, D-12277 Berlin
Books on Demand GmbH, Gutenbergring 53, D-22848 Norderstedt
Reha GmbH, Dudweiler Landstr. 99, D- 66123 Saarbrücken
ISBN: 978-3-8381-1016-5


**Imprint (only for USA, GB)**
Bibliographic information published by the Deutsche Nationalbibliothek: The Deutsche Nationalbibliothek lists this publication in the Deutsche Nationalbibliografie; detailed bibliographic data are available in the Internet at http://dnb.d-nb.de.
Any brand names and product names mentioned in this book are subject to trademark, brand or patent protection and are trademarks or registered trademarks of their respective holders. The use of brand names, product names, common names, trade names, product descriptions etc. even without
a particular marking in this works is in no way to be construed to mean that such names may be regarded as unrestricted in respect of trademark and brand protection legislation and could thus be used by anyone.


Publisher:
Südwestdeutscher Verlag für Hochschulschriften Aktiengesellschaft & Co. KG
Dudweiler Landstr. 99, 66123 Saarbrücken, Germany
Phone +49 681 37 20 271-1, Fax +49 681 37 20 271-0, Email: info@svh-verlag.de

Copyright © 2008 Südwestdeutscher Verlag für Hochschulschriften Aktiengesellschaft & Co. KG and licensors
All rights reserved. Saarbrücken 2008

Produced in USA and UK by:
Lightning Source Inc., 1246 Heil Quaker Blvd., La Vergne, TN 37086, USA
Lightning Source UK Ltd., Chapter House, Pitfield, Kiln Farm, Milton Keynes, MK11 3LW, GB
BookSurge, 7290 B. Investment Drive, North Charleston, SC 29418, USA
ISBN: 978-3-8381-1016-5

# Acknowledgments

This thesis was made possible by a collaboration between the Laboratory for Safety Analysis (LSA) and Huba Control AG in Würenlos. I would like to thank my doctoral supervisors first, Prof. Dr.Wolfgang Kroeger, director of LSA, Dr. Irene Eusgeld and Dr. Ralf Mock for their support and advice they have given me. I really appreciated the fruitful dialogues, the inspiring exchange, the constructive criticism concerning my work and the very efficient collaboration. Furthermore, my thank goes to Prof. Dr. Urs Meyer, my co-referee, mentor and emeritus director of the Institute for Automation at ETH Zurich, for the many lively discussions about technical but also off-topic issues. It has always been a pleasure and opened my mind for other aspects.

My special thanks are extended to the whole team at Huba Control AG, for their funding, trust, support and their helpfulness in every phase of my work. I am grateful for having had access to any information and the possibility to work in a well organized and equipped environment. Profound thanks are addressed to Mr. Roger Meier, CEO at Huba Control, who made this collaboration possible, Mr. Siegfried Lieckfeldt, head of production, for always having had time to discuss technical but also strategic questions concerning maintenance. Furthermore, I would also thanks to the rest of my core team at Huba, Mr. Juergen Budde, Mr. Michael Steimer, and Mr. Marc Voegeli for the sharing of technical expertise and the support in data collection, but also for the liberty in my time management. During my work I got in contact with almost every Huba employee. These numerous discussions gave me a priceless insight into an organization and were always a great pleasure. Thank you very much.

My deepest appreciation goes to Mr. Alois Stadler, head of quality management at Huba, for having been a great officemate and sharing his experience of many years of dealing with processes, organizations and people.

The greatest gratitude shall be exclusively reserved for my parents, who have always supported me during my studies in any possible way.

David Achermann, Zurich, October 2008

# Abstract

Globalization has increased the pressure on organizations and companies to operate in the most efficient and economic way. This tendency promotes that companies concentrate more and more on their core businesses, outsource less profitable departments and services to reduce costs. By contrast to earlier times, companies are highly specialized and have a low real net output ratio. For being able to provide the consumers with the right products, those companies have to collaborate with other suppliers and form large supply chains. An effect of large supply chains is the deficiency of high stocks and stockholding costs. This fact has lead to the rapid spread of Just-in-Time logistic concepts aimed minimizing stock by simultaneous high availability of products. Those concurring goals, minimizing stock by simultaneous high product availability, claim for high availability of the production systems in the way that an incoming order can immediately processed. Besides of design aspects and the quality of the production system, maintenance has a strong impact on production system availability.

In the last decades, there has been many attempts to create maintenance models for availability optimization. Most of them concentrated on the availability aspect only without incorporating further aspects as logistics and profitability of the overall system. However, production system operator's main intention is to optimize the profitability of the production system and not the availability of the production system. Thus, classic models, limited to represent and optimize maintenance strategies under the light of availability, fail. A novel approach, incorporating all financial impacting processes of and around a production system, is needed.

The proposed model is subdivided into three parts, maintenance and failure module, logistics module and production module. This subdivision provides easy maintainability and simple extendability. Within those modules, all cost-effective processes are modeled. Main part of the work lies in the extended maintenance and failure module that offers a representation of different maintenance strategies but also incorporates the effect of over-maintaining and failed maintenance (maintenance induced failures). Order release and seizing of the production system are modeled in the logistics part. Due to computational power limitation, it was not possible to run the simulation and the optimization with the fully developed production model. Thus, the production model was reduced to a black-box without higher degree of details.

This model was used to run optimizations concerning maximizing availability and profitability of the production system by varying maintenance strategies but also logistics factors. Those optimizations showed that there is a stringent connection between production system availability and logistic decision variables. This finding is a strong indicator that a joint optimization of logistics and maintenance strategies provides better results than optimizing those elements independently and highlights the need for the proposed sophisticated model.

Besides of the classic optimization criterion "availability", the overall profitability of the production system was investigated using a life-cycle approach coming from preinvestment analysis. Maintenance strategy was optimized over the whole lifetime of the production system.

It has been proved that a joint optimization of logistic and maintenance strategy is useful and that financial objective functions tend to be the better optimization criterion than production system availability.

# Zusammenfassung

Die Globalisierung der Märkte hat den Kostendruck auf die Unternehmen drastisch erhöht. Unternehmen wurden gezwungen, sich auf ihre Kernkompetenzen zu besinnen und haben begonnen, weniger profitable Unternehmensteile zu verkaufen und auszugliedern. Dadurch sank ihre Wertschöpfungstiefe und sie wurden gewzungen, sich mit anderen Unternehmen zu Wertschöpfungsketten zusammenzuschliessen. Mit der Anzahl der so verknüpften Unternehmen stieg auch die Menge der an Lager gehaltenen Güter, um die Versorgung der Endkonsumenten mit Produkten gewährleisten zu können. Dies hat zu einer schnellen Verbreitung von Just-in-Time Konzepten geführt, welche versuchen, diesem Widerspruch Rechnung zu tragen. Dieser Zielkonflikt bedingt eine hohe Verfügbarkeit der Produktionssysteme, damit eingehende Bestellungen umgehend bearbeiten werden können. Neben Design- und Konstruktionsaspekten sowie der Verarbeitungsqualität des Produktionssystems hat die Instandhaltung einen massgeblichen Einfluss auf die Verfügbarkeit.

In den letzten Jahrzehnten entstanden viele Instandhaltungsmodelle mit dem Ziel, die Verfügbarkeit zu optimieren. Die meisten konzentrierten sich dabei ausschliesslich auf Verfügbarkeitsaspekte ohne weitere Einflussfaktoren wie Logistik oder Rentabilität des gesamten Systems zu berücksichtigen. Daher versagen die klassischen Modelle, welche ausschliesslich die Verfügbarkeit im Fokus haben. Dies bedingt eine neue Vorgehensweise, welche alle finanzwirksamen Prozesse im und um das Produktionssystem abbilden kann.

Das vorgestellte Modell ist in drei Teile unterteilt, in ein Instandhaltungsmodul, in ein Logistikmodul und in einen Produktionsteil. Diese Unterteilung ermöglicht eine einfache Wartung und Erweiterung des Modells. Innerhalb der Module sind alle kostenwirksamen Prozesse abgebildet. Hauptteil der Arbeit liegt im erweiterten Instandhaltungs- und Fehlermodell, welches die Darstellung von unterschiedlichen Instandhaltungsstrategien ermöglicht aber auch die Effekte von Über- und fehlgeschlagener Instandhaltung abbilden kann. Die Einlastung von Aufträgen ist im Logistikmodul modelliert. Wegen limitierter Rechenkapazität war es nicht möglich, die Simulationen und Optimierungen mit dem detaillierten Produktionsmodell durchzuführen. Daher ist das Produktionsmodell nicht weiter detailliert worden. Mit dem Modell wurden Optimierungen durchgeführt, um die Verfügbarkeit, die Rentabilität des Produktionssystems aber auch logistische Faktoren zu maximieren. Die Optimierungen haben gezeigt, dass ein Zusammenhang zwischen der Verfügbarkeit von Produktionssystemen und deren Logistik besteht. Diese Erkenntnis deutet darauf hin, dass eine gemeinsame Optimierung von Logistik und Instandhaltungsstrategien bessere Ergebnisse liefert als eine separate Optimierung und bekräftigt den Bedarf an höher entwickelten Modellen wie das vorgestellte. Neben dem klassichen Optimierungskriterium "Verfügbarkeit" wurde die Rentabilität des Produktionssystems über seine ganze Lebensdauer analysiert. Dazu wurde eine Methodik aus der Investitionsrechnung verwendet, und die Instandhaltungsstrategie wurde über die ganze Lebensdauer optimiert. Es konnte gezeigt werden, dass eine gemeinsame Optimierung von Instandhaltungsstrategien und Logistikprozessen sinnvoll ist, und finanzielle Zielfunktionen als Optimierungsfunktion besser geeignet sind als die Verfügbarkeit des Produktionssystems.

# Abreviations

| | |
|---|---|
| BDD | Binary Decision Diagram |
| BSC | Balanced Scorecard |
| CBM | Condition-based Maintenance |
| CCF | Common Cause Failures |
| CF | Cash Flow |
| CFR | Constant Failure Rate |
| CM | Corrective Maintenance |
| CV[X] | Coefficient of Variance of Distribution X |
| DFR | Descending Failure Rate |
| DSS | Decision Support System |
| FSD | Functional Structure Diagram |
| FT | Fault Tree |
| ICT | Information and Communication Technology |
| IFR | Increasing Failure Rate |
| iid | Independent Identical Distributed Property |
| IRR | Internal Rate of Interest |
| JiT | Just-in-Time |
| MLE | Maximum Likelihood Estimator |
| MTBF | Mean Time Between Failures |
| MTTR | Mean Time To Repair |
| NPV | Net Present Value |
| OEE | Overall Equipment Effectiveness |
| p | Interest Rate |
| PM | Preventive Maintenance |
| PPC | Production Planning and Control |
| RCM | Reliability-centered Maintenance |
| TPM | Total Productive Maintenance |

# Contents

1 **Introduction**     1
    1.1 Background and Motivation . . . . . . . . . . . . . . . . . . . . . 1
    1.2 Problem Statement . . . . . . . . . . . . . . . . . . . . . . . . . . 4
        1.2.1 Objects of Investigation . . . . . . . . . . . . . . . . . . . 6
        1.2.2 Outline of the Thesis . . . . . . . . . . . . . . . . . . . . . 6
    1.3 Economic Importance of Maintenance . . . . . . . . . . . . . . . 7
        1.3.1 Investment Appraisal . . . . . . . . . . . . . . . . . . . . . 11
    1.4 Definitions and Terminology . . . . . . . . . . . . . . . . . . . . . 13
        1.4.1 Availability and Utilization . . . . . . . . . . . . . . . . . . 13
        1.4.2 Maintenance . . . . . . . . . . . . . . . . . . . . . . . . . . 14
        1.4.3 Definition of Terms related to Time . . . . . . . . . . . . . 18
        1.4.4 System and Component . . . . . . . . . . . . . . . . . . . 20
        1.4.5 Model, Modelling and Simulation . . . . . . . . . . . . . . 22
        1.4.6 Logistics . . . . . . . . . . . . . . . . . . . . . . . . . . . . 26

2 **Modeling of Availability Function**     29
    2.1 Static Methods . . . . . . . . . . . . . . . . . . . . . . . . . . . . . 29
        2.1.1 Boolean Model . . . . . . . . . . . . . . . . . . . . . . . . . 30
        2.1.2 Block Diagram . . . . . . . . . . . . . . . . . . . . . . . . . 31
        2.1.3 Fault Tree . . . . . . . . . . . . . . . . . . . . . . . . . . . . 31
        2.1.4 Binary Decision Diagram . . . . . . . . . . . . . . . . . . . 33
    2.2 Dynamic Methods . . . . . . . . . . . . . . . . . . . . . . . . . . . 34

|  |  | 2.2.1 | Alternating Renewal Process . . . . . . . . . . . . . . . . . . . . | 34 |
|---|---|---|---|---|

        2.2.2  Markov Chain Approach . . . . . . . . . . . . . . . . . . . . . . . 36

        2.2.3  Object Oriented Simulation . . . . . . . . . . . . . . . . . . . . . 38

**3 Connection between System Failure and Logistics**     **39**

    3.1  Matching of Output and Demand . . . . . . . . . . . . . . . . . . . . . . 39

        3.1.1  Production Output $O(t, T_i)$ . . . . . . . . . . . . . . . . . . . . . . 40

        3.1.2  Inventory . . . . . . . . . . . . . . . . . . . . . . . . . . . . . . . 42

        3.1.3  Demand . . . . . . . . . . . . . . . . . . . . . . . . . . . . . . . . 45

    3.2  Impact of System Failure on Service Level . . . . . . . . . . . . . . . . . 46

        3.2.1  Equation Derivation . . . . . . . . . . . . . . . . . . . . . . . . . 47

    3.3  Safety Factor $s$ and its Lower Boundary . . . . . . . . . . . . . . . . . . 48

        3.3.1  Minimal Safety Factor $s_{min}$ . . . . . . . . . . . . . . . . . . . . . 48

    3.4  Context of Forecasting Demand and Mission Availability . . . . . . . . . 50

        3.4.1  Planning Horizon $T_{PH}$ . . . . . . . . . . . . . . . . . . . . . . . 52

**4 Classic Maintenance Models and Failure Rate Shape**     **53**

    4.1  Classic Preventive Maintenance Models . . . . . . . . . . . . . . . . . . 54

        4.1.1  Failure Rate Preventive Maintenance Model . . . . . . . . . . . . 55

        4.1.2  Age Reduction Preventive Maintenance Model . . . . . . . . . . 56

        4.1.3  Hybrid Model . . . . . . . . . . . . . . . . . . . . . . . . . . . . . 57

    4.2  Shape of the Failure Rate . . . . . . . . . . . . . . . . . . . . . . . . . . 58

        4.2.1  Mixture Models . . . . . . . . . . . . . . . . . . . . . . . . . . . . 60

        4.2.2  Approximation of the Failure Rate $\lambda_{Failure}(t)$ . . . . . . . . . . . . 63

        4.2.3  Increasing Failure Rate . . . . . . . . . . . . . . . . . . . . . . . 65

        4.2.4  Constant Failure Rate . . . . . . . . . . . . . . . . . . . . . . . . 68

        4.2.5  Decreasing Failure Rate . . . . . . . . . . . . . . . . . . . . . . . 71

**5 Maintenance, Production and Logistic Model**     **75**

    5.1  Maintenance Model . . . . . . . . . . . . . . . . . . . . . . . . . . . . . 76

|  |  | 5.1.1 | Preventive Maintenance Model . . . . . . . . . . . . . . . . . . . . 79 |
|---|---|---|---|

       5.1.1  Preventive Maintenance Model . . . . . . . . . . . . . . . . . . . . 79  
       5.1.2  Repair Model . . . . . . . . . . . . . . . . . . . . . . . . . . . . . . 85  
       5.1.3  Failure Model . . . . . . . . . . . . . . . . . . . . . . . . . . . . . . 96  
       5.1.4  Cost Calculation . . . . . . . . . . . . . . . . . . . . . . . . . . . . 101  
  5.2  Logistic Model . . . . . . . . . . . . . . . . . . . . . . . . . . . . . . . . . 101  
       5.2.1  Service Level Determination . . . . . . . . . . . . . . . . . . . . 102  
       5.2.2  Production Calculation . . . . . . . . . . . . . . . . . . . . . . . . 105  
       5.2.3  Order Dispatch . . . . . . . . . . . . . . . . . . . . . . . . . . . . . 106  
  5.3  Production Model . . . . . . . . . . . . . . . . . . . . . . . . . . . . . . . 110  
       5.3.1  Manufacturing . . . . . . . . . . . . . . . . . . . . . . . . . . . . . 111  
       5.3.2  Production Calculation . . . . . . . . . . . . . . . . . . . . . . . . 112  
  5.4  Top-Level Environment . . . . . . . . . . . . . . . . . . . . . . . . . . . . 113  
       5.4.1  Discounted Cash Flow . . . . . . . . . . . . . . . . . . . . . . . . 113  

## 6 Simulation and System Optimization       117

  6.1  Goals of the Simulation . . . . . . . . . . . . . . . . . . . . . . . . . . . . 117  
  6.2  Design of Experiments . . . . . . . . . . . . . . . . . . . . . . . . . . . . 118  
  6.3  Simulation Run of the Default Production System (Exp. 1) . . . . . . . . 120  
       6.3.1  Model Accuracy . . . . . . . . . . . . . . . . . . . . . . . . . . . . 120  
       6.3.2  Availability and Service Level . . . . . . . . . . . . . . . . . . . . 121  
  6.4  Production System without Preventive Maintenance (Exp. 2) . . . . . . 124  
  6.5  Sensitivity Analysis of the Safety Factor (Exp. 3) . . . . . . . . . . . . . 125  
       6.5.1  Impact of the Safety Factor on System Availability . . . . . . . . 126  
       6.5.2  Effect of the Safety Factor on the Service Level . . . . . . . . . . 126  
       6.5.3  Influence of the Safety Factor on the Cash Flow and Discounted Cash Flow . . . . . . . . . . . . . . . . . . . . . . . . . . . . . . . . 127  
  6.6  Optimization . . . . . . . . . . . . . . . . . . . . . . . . . . . . . . . . . . 127  
       6.6.1  Optimizing Techniques . . . . . . . . . . . . . . . . . . . . . . . . 128  
       6.6.2  Optimization Stop Conditions . . . . . . . . . . . . . . . . . . . . 132

      6.6.3   Optimization Criteria . . . . . . . . . . . . . . . . . . . . . . . . . 133

      6.6.4   Maximizing $A_{SS}(t)$ (Opt. 1) . . . . . . . . . . . . . . . . . . . . 134

      6.6.5   Maximizing $CF(t)$ (Opt. 2) . . . . . . . . . . . . . . . . . . . . . 135

      6.6.6   Maximizing $SumDCF(t)$ (Opt. 3) . . . . . . . . . . . . . . . . . . 135

      6.6.7   Comparison of Optimized Systems with the Default System . . . . 137

  6.7  Recommendation . . . . . . . . . . . . . . . . . . . . . . . . . . . . . . . 139

# 7  Conclusions and Outlook  143

# List of Tables  148

# List of Figures  149

# List of Abbreviations  154

# Bibliography  155

# A  Novel Approach of Availability Modelling  171

  A.1  Approximation Techniques for dealing with Simultaneous Failures . . . . . 172

      A.1.1   Inclusion-Exclusion Principle . . . . . . . . . . . . . . . . . . . . . 173

      A.1.2   Rare Event Approximation . . . . . . . . . . . . . . . . . . . . . . 174

      A.1.3   Structure Function Methods . . . . . . . . . . . . . . . . . . . . . 175

      A.1.4   Approaches using Canonical Form . . . . . . . . . . . . . . . . . . 177

  A.2  Novel Approach . . . . . . . . . . . . . . . . . . . . . . . . . . . . . . . . 179

  A.3  Laplace Transformation . . . . . . . . . . . . . . . . . . . . . . . . . . . . 180

      A.3.1   Serial System . . . . . . . . . . . . . . . . . . . . . . . . . . . . . . 181

  A.4  Expansion to other Subsystem-Types . . . . . . . . . . . . . . . . . . . . 182

      A.4.1   Parallel System . . . . . . . . . . . . . . . . . . . . . . . . . . . . 183

      A.4.2   k-out-of-n System . . . . . . . . . . . . . . . . . . . . . . . . . . . 183

  A.5  Assembling Layouts . . . . . . . . . . . . . . . . . . . . . . . . . . . . . . 183

      A.5.1   Tree Assembling . . . . . . . . . . . . . . . . . . . . . . . . . . . . 185

|       | A.5.2 Network Assembling | 186 |
|---|---|---|
|       | A.5.3 Block Assembling | 186 |
| A.6 | Buffers | 187 |
|       | A.6.1 Buffer Modelling | 188 |
| A.7 | Example | 190 |
| A.8 | Consequences of the Improved Modelling Approach | 191 |

## B  Queueing Theory   193

## C  Demand Forecast   197

- C.1 Qualitative Methods ... 197
- C.2 Quantitative Methods ... 198
  - C.2.1 Causal Methods ... 199
  - C.2.2 Time Series Extrapolation ... 199
  - C.2.3 Elementary Technique ... 200
  - C.2.4 Moving Average Forecast ... 200
  - C.2.5 First-Order Exponential Smoothing Forecast ... 200
  - C.2.6 Trigg and Leach Adaptive Smoothing Technique ... 201

## D  Maintenance Strategy Optimization Procedures   203

- D.1 Reliability-Centered Maintenance (RCM) ... 203
- D.2 Total Productive Maintenance (TPM) ... 206

# Chapter 1

# Introduction

The first chapter shall give a brief introduction into subject of maintenance, its associated areas of conflict and trends in the industry. In addition to present the most important maintenance strategies and maintenance selection procedures, their impact on industry and company level is discussed.

## 1.1 BACKGROUND AND MOTIVATION

Companies in the field of manufacturing industries are more and more interlinked in supply chains, which have grown in width and depth. Increasing pressure on costs and response time are the main driving forces of this development and have lead to more and more diversified and specialized suppliers, and therefore to an increasing amount of participants in a supply chain. Those tendencies promoted the spread of Just-in-Time (JiT) concepts for production planning and control. In comparison with resource-oriented conceptions as Loor, Corma or Kanban, JiT is aimed at increasing the potential to short lead times and minimizing stock and work in process (see [Schoensleben, 2002]). Maximal utilization of the production system is not strived. However, since JiT minimizes inventory and work in process, manufacturing costs can be reduced. Shortened lead times move the stocking level towards lower levels. Thus, a larger part of the value added chain is within the lead time required by the customer. Stocking cheaper intermediate goods reduces storage costs. Not only inventory costs are decreased but also potential costs due to inaccurate forecasting are minimized. Whereas costs are reduced in JiT, the probability of being non-deliverable increases due to diminished inventory.
Providing a continuous flow of final products to the customer requires high service lev-

els of the company. Service level is understood as the ratio between orders delivered on time divided by all orders (see Definition 1.4.6). High service level $S_L(t, k)$ may only be achieved by high inventory stock on finalized product level or high availability of production system in combination with short production times. Stockkeeping of products in only beneficial in the case of standard products and continuous or regular demand. As long as the variability of the products is low, stockkeeping is an admissible solution to equalize variance in production system availability. However, in the situation of lumpy demand or customized products, products cannot be held on stock, since warehousing would result in immense costs. By consequence, companies are compelled to increase and secure availability of their production system and optimized maintenance strategies are an approach to meet these requirements.

This research project has a partner in manufacturing industries. The industry partner is facing the fact that several clients claim a continuous delivery of products. The company operates only one facility and manufactures pressure switches and sensors for applications in the automotive industries and in the heating, ventilation and air-conditioning sector. Most of those products are manufactured on non-redundant plants. Increasing pressure on costs and demand for shorter time-to-delivery, coupled with severe penalties for default in delivery, force the company to invest in production system availability and sophisticated maintenance strategies. A high readiness for delivery claims high available production system; maintenance has a strong impact production system availability. Figure 1.1 encompasses the area of conflict of the company.

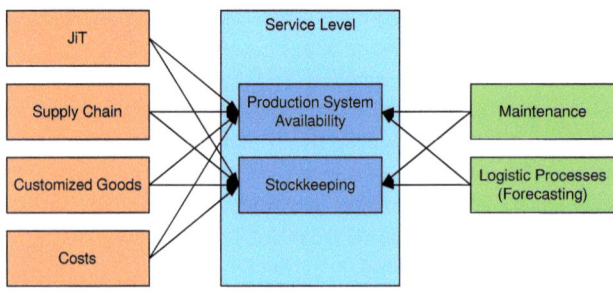

**Figure 1.1:** *Area of Conflict in the Field of Maintenance and Logistics*

Production planning and control and concepts of production are of importance and set additional constraints to the required production system availability. By optimizing the overall profitability of a production system, dependence between maintenance strate-

## 1.1. Background and Motivation

gies, logistics processes, production planning and control and concept of production must be taken into consideration. Analytical approaches and methodologies are not applicable to such a complex problem because of interdependence and interplay of all elements in the system. A schematic depiction of the parameters and their influences on the target figures production system availability $A_{SS}(t)$, cash flow $CF(t)$, discounted cash flow $DCF(t)$ and service level $S_L(t)(t, k)$ are shown in figure 1.2. The parameters printed in red are the decision or control factors.

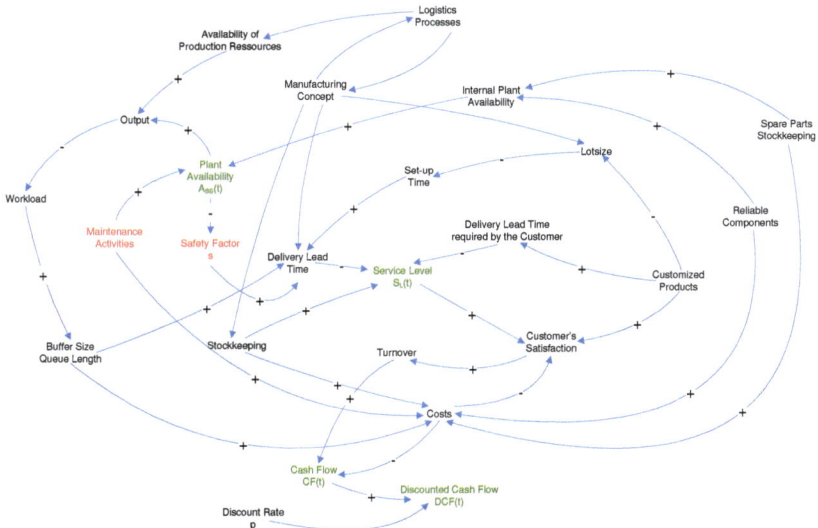

**Figure 1.2:** *Cause and Effect Diagram of Maintenance*

Off-the-shelf and best-practice methods to select maintenance strategies, as well as spare parts stockkeeping, are widely spread in manufacturing industries; they are mainly based on experience and production system manufacturer proposals. Improvements are normally done step-wise in a trial and error manner without accounting for cost-effectiveness. A strict line of action is absent, which often originates from missing objectives and monitoring tools. Furthermore, most maintenance strategy optimization techniques are addressed to increase system availability solitarily without accounting for customer need as minimal service level or cost-effectiveness of the whole production system. However, maintenance actions should always be aimed at optimizing the profitability of operating a production system. This encompasses all expenditures but also demand and associated turnover. Since supplying demand requires a minimal production system availability, optimizing system availability (optimizing the maintenance strategy) needs an involvement of demand. Optimizing the maintenance strategy under consideration of demand is a complex task and needs simulation support for decision making.

## 1.2 Problem Statement

Although the economic contribution of maintenance to the company profitability is beyond dispute (see 1.3), many companies regard maintenance and the maintenance department as expense factor only. Maintenance is a significant cost factor in many companies and is under constant pressure of cost reduction. Among others, the tendency to highlight costs and disregarding the benefit of maintenance is fostered by the difficulties to rate and estimate the contribution of maintenance to the company's profit. Even though many rating and optimization approaches (e.g. Reliability Centered Maintenance (RCM) [Moubray, 1991] and Total Productive Maintenance (TPM) [Nakajima, 1988]) have been developed, they still lack of reliable quantitative measurands and impede a cost-benefit consideration between different maintenance strategies but also among other investment ventures. Since availability has a substantial impact on the profitability of a production system, sophisticated and state of the art maintenance strategy optimizations should always be based on models being capable to represent this factor. The most widespread methods applied to model availability are Petri-Nets and Markov-Chains (an encompassing outlook about those techniques and their deficiencies is given in chapter 2). However, maintenance strategy optimization approaches in practice make hardly use of those quantitative models (see appendix D).

Some of the reasons for their spare application in maintenance optimization are:

**Cumbersome modelling** Transformation of the problem into a model is difficult and hardly intuitive. Every model requires its own modelling language with its own symbolism and syntax. Elements and dynamics of the real system have to be translated into the modelling language.

**Inefficient modelling techniques** Modelling is time-consuming. The literal modelling process can hardly be accelerated since the reuse of those models or some parts of the models is practically impossible. Furthermore, there is a trade-off between tools with a very limited set of functionality but providing efficient and effective

## 1.2. Problem Statement

modelling and sophisticated methodologies with encompassing functionality that suffer from insufficient efficiency.

**Limited extendability** Integration of further system aspects as demand or logistic strategies cause mostly a complete re-modelling and re-programming of the whole model and can hardly be implemented.

**Inadequate modelling of preventive maintenance impact on availability** Only a few models exist that are able to represent impact of preventive maintenance on system availability. They model the influence with a failure rate modification, meaning that a preventive maintenance reduces the instantaneous system failure rate. The greatest deficiency of those models is their inability to represent the effect of over-maintaining and maintenance induced failures. A system may be exhaustively maintained which can end in a higher failure-proneness and decreased system availability.

**Loss of analytical solvability** Simple modelling tools allow to analytically solve the problem whereas most advanced models can solely be analyzed by means of simulation. Not only solution finding but also optimization becomes more demanding. Within this framework, questions about the reliability and sensitivity of simulation results appear which refer to the difficultiy of local and global maxima.

**Computational power** All advanced availability models need to be implemented into a simulation environment to get quantitative results. Evaluation of those models refers to solve Boolean equations. The difficulty about Boolean equations is their tendency of combinational term explosion. Larger and more complex systems require enormous computational power for equation evaluation or computation may even be impossible.

Within the last few years, JiT-logistics and the pressure on costs and delivery on time have dramatically gained importance and have companies urged to optimize their service level. With respect to maintenance strategy optimization, this development has impacted maintenance objectives as well. It indicates a tendency away from optimizing system availability towards optimizing the service level and the overall profitability of the production system. Since this is the case, it is arguable why logistics and maintenance objectives are separately optimized and if a joint optimization would not provide better results. Therefore, an integration of maintenance, availability and logistics model into one, all-embracing model is required.

## 1.2.1 Objects of Investigation

As before, 1.1, the objective of this work is to investigate the interrelationships between maintenance and logistic requirements and how maintenance may contribute to logistics objectives, as service level and fill rate. In an analytical approach, the fundamental mathematical relations between service level, fill rate and system availability will be derived. These equations are used to deduce minimal logistics and maintenance requirements and should provide a proof of evidence of the multiple interrelations between logistics and maintenance.

Moreover, some of the difficulties mentioned above should be overcome by applying state-of-the-art modeling and simulation techniques. An overview about existing approaches to model system availability is given. Moreover, an encompassing model incorporating the impact of maintenance on logistics measurands and the system availability shall be elaborated to make a simultaneous optimization of logistics, maintenance and financial aspects possible. The model is implemented in an agent-based simulation environment to run simulations and optimizations. Besides optimizing availability of a production system by applying different maintenance strategies, the impact of a chosen maintenance strategy on the overall profitability of a production system is investigated.

## 1.2.2 Outline of the Thesis

After the introductory chapter, the remainder of the work is organized as following:

**Chapter 2** is subdivided into three sections giving a comprehensive outlook about static, dynamic and approximation modeling techniques applied in reliability engineering. Approximation techniques are mainly addressed to diminish the tendency of combinational term explosion. The objective of this chapter is to discuss strengths and weaknesses of the different modeling methods and to highlight the difficulties in their application.

**Chapter 3** introduces and discusses the relationship between maintenance and logistics. Equations for service level and fill rate will be deduced and linked with system downtimes and system availability. Those equations provide the strict interrelation between maintenance and logistics. Hereafter, it will be shown that lower boundaries for production planning and control exist coming from demand forecasting and mission availability requirements.

**Chapter 4** presents classic maintenance models incorporating the effect of preventive maintenance on the failure rate. The concept of failure rates subdivided into maintainable and non-maintainable failure rates is presented. Some of the described models assign different quality levels to the maintenance activities which enable to model maintenance induced failures. The second half of the chapter is concerned with the failure rate shape evaluation.

**Chapter 5** constitutes the main part of this work by introducing the extended, encompassing maintenance model. The model is divided into three sections, a maintenance, a production, and a logistics model which are explained individually. Within the maintenance model, the ideas of quality levels, segmented failure rate and the impact of maintenance on the failure rate are realized. Whenever a failure occurs or a maintenance activity is performed in the maintenance model, a message is sent to the production model and the production process is interrupted. The literal production process is represented in the production model which is assumed to be a black-box. Production orders are released from the logistics model and are routed to the production model. When the production orders are processed in the production order, they are routed back in the logistics model. In principle, the logistics model is mainly used for service level and fill rate calculation.

**Chapter 6** discusses the simulation and optimizing results of the model described in chapter 6. Simulation and optimization are performed on an agent-based simulation engine called $AnyLogic^{©}$.

**Chapter 7** summarizes the results and identifies interesting fields for future works.

## 1.3 ECONOMIC IMPORTANCE OF MAINTENANCE

Economic impact of maintenance has gained importance and is still getting stronger. [Eichler, 1990] speaks about that 10 to 30% of all employees work in the maintenance sector; in highly automated industries over 60%. This amount will increase due to higher automated production systems and, therefore, less operationally working personal. Raised production system complexity and spare parts diversity have strengthened the impact of maintenance and have lead to higher requirements regarding system availability and reliability [Neuhaus, 2007]. Contribution and value-added of maintenance support company's competitiveness by reducing overall costs and optimizing production system availability.

Production systems have grown in complexity and demand for highly skilled, well educated and expensive maintenance staff. Surveys on maintenance costs in Scandinavian companies show that 32% of its maintenance budget is spent on salary and wages, 32% on spare parts, and 31% on external services (mainly wage payments for third parties). Hence, increase in wage payments due to soared amount of maintenance employees and their salaries (advanced qualification requirements) have constantly risen maintenance costs between 10 to 15% per year in the US [Luxhoj et al., 1997]).

Not only maintenance costs have soared but also costs for lost production showed an over-proportional growth. The relation between cost for lost production and maintenance costs are 4 to 1 on average and can increase up to 15 to 1 in some cases [Luxhoj et al., 1997]. It is expected that this relation will augment in the future and that maintenance strategies will be aimed at preventing production interruptions.

The overall direct maintenance costs (wages and salaries, material, spare parts, legal consequences, stock-keeping costs, etc.) are estimated to be in the range of 1'500 billion Euro and the indirect costs (machine failures, quality losses, image losses, etc.) are around 7'500 billion Euro in the European Union for the year 2007 [Podratz, 2007] (see Figure 1.3). With a turnover of over 1'500 billion Euro, maintenance sector is one of most important economic sectors in Europe and, in all probability, in any industrialized economy contributing with around 10% to the Gross Domestic Product (GDP) [Warnecke, 1992].

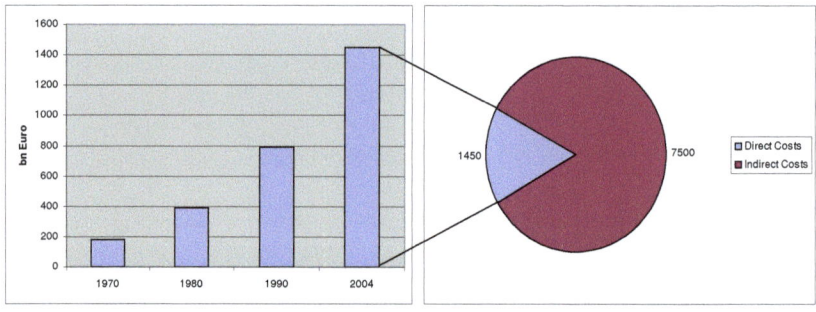

**Figure 1.3:** *Direct and Indirect Maintenance Costs*

Moreover, maintenance influences several other factors which directly or indirectly con-

## 1.3. Economic Importance of Maintenance

tribute to company's profit. With respect to a survey [Bandow, 2006] on 240 companies in Europe these factors are:

- Overall costs reduction
- Improvement of job and plant safety
- Achievement of production objectives
- Achievement of quality objectives
- Minimizing resources
- Compliance of environmental protection obligations
- Minimizing negative effects on environment
- Increasing working conditions

Age distribution and current replacement value of the capital assets mainly impact maintenance costs and expenditures. Ageing of production systems and their required maintenance intensity share a common trend, maintenance costs raise with increasing age of the production system [Warnecke, 1992]. Another correlation is described by the indicator *degree of maintenance* that sets maintenance costs in relation to current replacement value of a production system. For complex production system, this indicator can easily reach values between 3 and 10%, representing maintenance costs in the range of 3 and 10% of the current replacement value of the production system [Warnecke, 1992]. Fast development in highly automated and networked production systems in the recent years suggests that this ratio has increased and (see [Alcade, 2000] who cites a study that shows that maintenance costs have exposed a larger increase than production costs; a strong indicator for a drift towards more intensified maintenance activities). This high degree of maintenance may cause costs that can reach almost 5% of the company's turnover (compare with [Luxhoj et al., 1997] and [Kastner and Dankl, 1992]).

Integration of information and communication technology (ICT) in production systems to allow data exchange with production planning and control (PPC) and accounting systems expose the tendencies towards more expensive and complex production systems with higher demand of maintenance. Whereas labor production costs have sunken the last years due to the substitution of work by capital, this quota is increasing in maintenance (also partially impacted by the increased wage level of the maintenance labor).

It can be concluded that maintenance costs have dramatically gained importance and are one of the main cost-drivers, in particular in highly-automated industries, and that optimizing maintenance strategies will be one of the major challenges for companies in the manufacturing industry in the next years.

Purchasing, operating and maintaining a production system induce expenditures and costs. Those expenditures front proceeds raising the question has to combine factors as capital costs, operating, costs, maintenance costs, performance, or availability maximizing the production system profitability. Analysis of those factors is rendered difficult by the high uncertainty of most costs and revenues, and the multitude of complex interdependencies. This matter of fact is, among others, the reason why production system acquisition appraisals are dominated by factors about which good estimates do exist (those factors are capital costs and production system performance, generally). However, those factors often do not represent the main cost drivers. In some cases, overall maintenance (and operation) costs can be considerably higher than capital costs (see [Kelly, 1997]) and may strongly influence production system profitability. This tendency has fostered the awareness, that life-cycle considerations will provide better results concerning optimization of production system cost structure.

Life-cycle costing incorporates all phases of the lifetime of a production system, from the point in time where the need for a production systems becomes obvious until disposal. From the perspective of life-cycle costing, the most cost-efficient investment is the one causing the least possible costs. The major drawback about this accounting method is its limitation to solely incorporate costs without accounting for the revenues. Thus, any profitability deliberation cannot be considered with life-cycle costing.

Not only the appraisal of future maintenance costs, but also the influence of a chosen maintenance strategy on the production system is subject to complex evaluations. There have been many attempts to quantify the benefit and impact of different maintenance strategies on the economic and performance result of production systems. Most of them referred to a modification of the production system failure rate due to maintenance (see [Lie and Chun, 1986], [Nagakawa, 1986], [Nagakawa, 1988], [Canfield, 1986], [Malik, 1979]). Then, costs for maintenance are contrasted with the gain of production system availability resulting in higher service level and increased system output.

Classical methods, as described in [Eichler, 1990], [Warnecke, 1992] or [Adam, 1989], follow the traditional cost accounting. All of those methods expose the deficiency that they abstract the time value of money. Time value of money results from the concept of

## 1.3. Economic Importance of Maintenance

interest that an investor prefers to receive a payment of a fixed amount of money today, rather than the same amount in the future [Bodie and Merton, 2003]. Money received today can be used for further investment, that earns interest and compound interest, or for consumption purposes. On the other hand, obligations with maturity in the future are lower than obligations one has to pay back now (see Figure 1.4). This concept must be taken into considerations when dealing with future cash-flows.

**Definition**     **1**     *Cash flow $CF(t)[sFr.]$ is referring to the difference of received and spent amount of cash by a business during a predefined period in time and stands for the increase of cash from operating activities [Seiler, 2000].*

Future cash-flows should be discounted back to a reference point in time to provide a common basis for profitability comparisons of different investment and maintenance strategies. A maintenance strategy is an investment. Then, the well-known investment appraisal techniques can be applied to find the optimal investment, respectively the optimal maintenance strategy (see subsection 1.3.1).

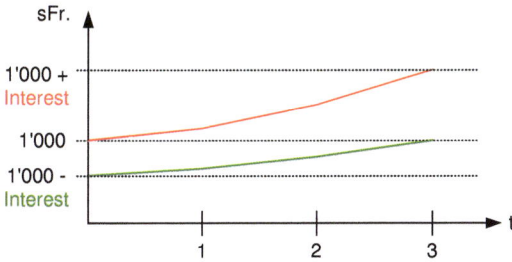

**Figure 1.4:** *Time Value of Money*

### 1.3.1 Investment Appraisal

Investment appraisal is a tool for estimating the profitability of an investment and is focused on the quantitative aspects of an investment project. It is apparent, that incorporating only quantitative effects can never give an encompassing impression about the quality of an investment. Qualitative facets are at least as important but should not be the essence of this explanations (see [Seiler, 2000] for a detailed discussion about qualitative aspects in investments). Furthermore, investment appraisal offers a toolbox for contrasting different possibilities and helps selecting the best one. A short general introduction into investment appraisal methods is given.

In principle, approaches using interest and compound interests and static methods neglecting the present value of money can be differentiated. Their most important exponents are payback-calculation as part of the static methods and net present value (NPV) and internal rate of interest method (IRR) for approaches using interest and compound interests (dynamic methods). Since Watson et. al. [Watson and Head, 2007] noticed that NPV is the method of choice, at least in larger companies, only this method will be presented.
The basic concept of all dynamic investment appraisals incorporates only the future cash flows $CF(t)$ caused by the investment. Those annual cash flows are discounted back to a reference point with a discount rate $p$ and are called discounted cash-flows $DCF(t)[sFr.]$. The sum of all $DCF(t)$'s is the time value of money of all future cash flows and is denoted as $SumDCF(t)[sFr.]$.
The NPV-method contrasts this sum of discounted cash flows with the initial investment. This method allows a calculation of the surplus of the investment, thus the residual amount of money after amortization and calculatory interest. Net present value $(NPV)$ of an investment denotes the surplus between the initial investment $G_0$ and the sum of all discounted cash flows $SumDCF(t)$. [Seiler, 2000].

$$NPV(t) = -G_0 + SumDCF(t) \tag{1.1}$$

$$SumDCF(t) = \sum_{t=1}^{n} \frac{CF(t)}{(1+p)^t} \tag{1.2}$$

$p$ : Interest rate
$G_0[sFr.]$ : Initial investment
$CF(t)[sFr.]$ : Cash flow at the end of year $t$

As long as the $NPV(t)$ of an investment is positive or zero, meaning that the sum of all discounted cash flows $SumDCF(t)$ is larger or equal to the amortization and calculatory interest, the investment is profitable. Since the initial investment is independent from the maintenance strategy, $G_0$ can be set to zero for maintenance optimization purpose and the optimizing factor is $SumDCF(t)$.
Thus, the maintenance strategy with the highest $SumDCF(t)$ is the most profitable. In

## 1.4. Definitions and Terminology 13

comparison with the classic optimization approach, $SumDCF(t)$ takes all future cash flows into account and optimizes the investment profitability over the whole production system lifetime.
A main characteristic of the $SumDCF(t)$-calculation is its tendency to weight early returns stronger than later cash flows, particularly in combination with high interest rates. This may lead to the effect that an increase of maintenance costs due to ageing of the production system is partially or even fully compensated by the discounting.

## 1.4 DEFINITIONS AND TERMINOLOGY

In this section, the most important terms are defined and grouped around umbrella terms to establish a universe of terminology.

### 1.4.1 Availability and Utilization

If the effectiveness and profitability of a maintenance strategy shall be investigated in both, a qualitative and quantitative manner, performance criteria need to be defined. Among others, availability of a production system can provide some quantitative information about the performance of a maintenance strategy. Availability indicates the readiness of a system and is defined as:

**Definition 2** *Availability* $A_{SS}(t)$ *is the probability that a system is in operational state at a certain point in time $t$ [DIN-40041, 1990].*

Thus, availability is a question of meeting specifications as performance, quality or cost requirements, for example. Whether a system is available depends on the predefined threshold boundaries concerning specifications to be complied with. These requirements incorporate mostly performance, quality, and cost specifications, but are not only limited to technical issues. When a system falls short of one of these operating thresholds, the system is unavailable. However, this definition implies that availability $A_{SS}(t)$ is a *transient availability* providing information about only one specific point in time $t$. For theoretical consideration this might be sufficient, but fails when it comes to real-world applications. Practitioners are rather interested in getting an availability estimate about a period in time than about a point probability. The average availability during a period in time $T_i$ is called *mission availability* $A_{Mission}(t, T_i)$.

**Definition**     **3**    *Mission availability* $A_{Mission}(t, T_i)$ *is defined as:*

$$A_{Mission}(t, T_i) = \frac{1}{T_i} \int_t^{t+T_i} A_{SS}(t) dt$$

### 1.4.2 Maintenance

Breakdowns and holdups in production systems can seriously impacting system availability and usability, both putting profitability of a production system at risk. Idle production systems cause a negative shift regarding the ratio between fixed costs to production output. In combination with the reduced production output due to breakdowns, this has a double negative effect of the cost-effectiveness of the production system [Seiler, 2000]. Moreover, sophisticated production systems often need significant start-up time after an interruption. During this time, scrap or goods of minor quality are manufactured that either cannot be sold or only at reduced prices. Thus, efficient operation of a production system claims only few interruptions and fast recovery from breakdown.

**Definition**     **4**    *Maintenance depicts the entity of all technical, technological, organisational, and economic actions to delay wearout and/or recovery of functional capability, including technical safety, of a technical system [DIN-31051, 2001]. The umbrella term maintenance can be subdivided as shown in Figure 1.5.*

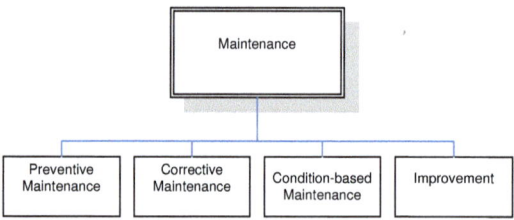

**Figure 1.5:** *Classification of Maintenance*

## 1.4. Definitions and Terminology

Wear-out, ageing and deteriorating processes have a negative impact on the functional capability of technical system and the system may fail to fulfill its assigned functionality. Maintenance is designated to counteract those effects in an economic way. Economic use of maintenance is dedicated to a smart combination of the four distinctive parts of maintenance, preventive, corrective, condition-based maintenance, and improvement, and to sequentially arrange them on the time axis. The following explanations are inspired by [Warnecke et al., 1996], [Eichler, 1990], [Birolini, 2007], [Moubray, 1991], [Wunderlich, 2005] and [Kelly, 1997].

**Corrective maintenance (CM)** is initiated after a failure occurs and is intended to reset system into a failure-free state [DIN-13306, 2001]. Often, corrective maintenance is named repair or restoration and involves the actions repair and replacement of failed components. This type of maintenance can be applied in systems where failures do not cause costly and dangerous situations, for components with constant failure rate (expose purely stochastic failures), or for systems with built-in redundancy. Benefit of corrective maintenance is the maximum exploitation of the wear-out reserve of the components. However, CM is, in some form, an integrative part in any maintenance strategy, since unplanned breakdowns can never be excluded.

**Preventive maintenance (PM)** encompasses all activities geared towards reducing or preventing deteriorating tendencies by anticipating possible future failures. It includes partial or complete overhauls, oil changes, etc. with the focus to prevent the system from failure before it actually occurs. Preventive maintenance makes sense when the failure rate of a component increases in time, when the costs for preventive maintenance are lower than the overall costs of a breakdown strategy (CM), or a breakdown could lead to severe accidents. Since PM reduces unplanned production holdups due to stochastic breakdowns and CM, production planning and controlling (PPC) is simplified. In comparison to CM activities, PM tasks can properly be prepared (spare parts are available, maintenance crew can be trained, etc.). Preparation reduces downtime and the tendency of maintenance induced failures. Although preventive maintenance is designated to prevent the system from failure, some failures may still occur. Those stochastic failures are covered with corrective actions. Thus, a preventive maintenance strategy incorporates always reactive (CM) and proactive (PM) tasks.

Just as in any maintenance activity, there are risks of human errors or equipment failures when performing preventive maintenance (see subsection 5.1.2.2). Exhaustive use of preventive maintenance may lead to maintenance induced failures in the way that the preventive maintained system is less available than the unmaintained system. Another deficiency is related to the fact, that the wear-out reserve of preventively replaced components is not fully utilized. However, in situations where failures may end in catastrophic consequences or enormous losses this may be the only feasible option if condition-based maintenance is not applicable.

**Condition-based maintenance (CBM)** incorporates inspections of the system in pre-determined intervals to determine system condition. Depending on the outcome of an inspection, either a preventive or no maintenance activity is performed. Thus, CBM is a variety of a PM strategy with the difference, that the triggering event to perform a preventive maintenance activity is the expiring of a maintenance interval in the PM case, respectively the result of an inspection in CBM. It aims at maintaining the right components, with the adequate maintenance activity at the best point in time by using real-time data to prioritize and allocate maintenance resources. Apparently, CBM is only applicable when wear-out reserve is measurable. In 2001, a joint effort by the industry and the US Navy was undertaken to develop a standardized approach of information exchange within the CBM community. They anticipated that a unified framework for exchanging information would increase efficiency and decrease costs for CBM-systems [Deventer, 2006]. The OSA-CBM specifications are the result of this endeavor, providing a standard architecture for moving information across the community and a guideline for implementing CBM-systems. Explanations below follow the idea of OSA-CBM (see Figure 1.6).

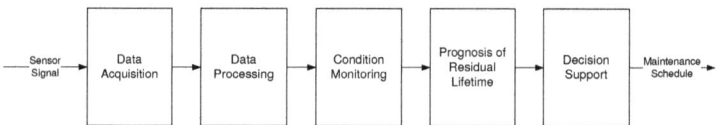

**Figure 1.6:** *OSA-CBM Framework*

Acquisition of data is done by observing the state and condition of the production system with monitoring tools and devices. Among others, some of the monitoring tools are:

- Vibration monitoring
- Lubrication monitoring

## 1.4. Definitions and Terminology

- Thermography
- Acoustic sound source localization
- Non-destructive thickness measuring with ultrasonic

Inspection outcomes have to be processed to detect, isolate and identify a fault [Fabricius, 2003]. A fault is a condition deviated from the normal characteristics that can lead to a failure (inability to provide a certain service). For this reason, a sound definition of the "normal" condition as well as upper and lower boundaries, when this state is left, are required. Fault detection methods make mostly use of a common strategy, they compare present data with some reference data. If the reference data are the outcome of an exemplary representation of the real system, the fault detection method is called model-based [Gertler, 1998]. This representation is a simplified model of the real system addressed to imitate its characteristics.

Mainly, two distinctive kind of models can be identified, analytical models and models related to methods used in machine-learning. Whereas analytical models are limited to represent linear characteristics, modern modelling techniques base on artificial intelligence, as neuronal networks [MacKay, 2003], Bayesian (beliefs) network or support vector machines [Bishop, 2006], and are also capable to cover nonlinearities and complex inter-dependabilities. In the case, where reference data are data of the real, faultless system, the technique is called model-free method. The outcome of both model types serves as an indicator for the actual condition of the system/components and is used for diagnosis. Depending on the diagnosis, expected residual lifetime of the components are evaluated and combined with experiences to decide whether a maintenance activity should be performed or not. Contemporary CBM systems make use of sophisticated software to manage information about production system history and to estimate efficiency of different maintenance tasks. Those systems have an integrated decision support tool to provide recommendations about what to do best concerning minimizing spare parts cost, system downtime or maintenance time. Decision support systems (DSS) belong to computer-based information systems that support the act of decision making. In this case, decision making is understood as making a choice between alternative maintenance tasks on the basis of the diagnosis results. A DSS may contribute in this process by supporting the estimation, the evaluation and the comparison of alternatives by providing deeper insight into this unstructured problem. Since the resulting schedule of maintenance activities is based on the expected residual lifetime of some components, CBM is also called predictive maintenance.

CBM offers the opportunity to run a production system in an optimal mode, optimal with respect to the chosen objective function, as costs or system reliability. Concerning the exploitation of wear-out reserve, CBM is somewhere between CM and PM. A condition-based maintained system leads to higher system reliability, increased availability and lower production costs by lower utilization of resources in comparison with PM and CM. Since environmental issues become more and more important, an economic use of scarce resources may be one of the essential unique selling propositions in the future.

Introducing CBM is a cost-intensive action. Required instrumentation of equipment to detect and monitor system condition may cause high costs, especially retrofitting of already existing production systems. Not only the monetary aspects may cause some difficulties but also the technical side can be complicated. Although the condition of some components can easily be monitored by measuring and recording simple values (e.g. temperature, vibration), transformation of this knowledge into counteracting maintenance tasks can be awkward.

An encompassing listing of benefits and disadvantages of CM, PM and CBM can be found in [Fabricius, 2003].

The assignment of specific maintenance tasks to distinguished technical systems in workscope and time is called *maintenance strategy*.

**Definition    5**    *A **maintenance strategy** defines type, content, and temporal sequence of maintenance tasks for a technical system.*

Strategy contains a methodical advancement in which tasks are brought into an efficient and plausible sequence. Efficiency is aligned according to some objectives the strategy is geared towards. A maintenance strategy is always aimed at a specific goal. Those goals can incorporate e.g. monetary guidelines, safety or security claims, or minimal system availability.

### 1.4.3    Definition of Terms related to Time

Figure 1.7 defines and depicts the relation between the most important different temporal measurands. Those definitions allow a specification of the sources of time losses.

## 1.4. Definitions and Terminology

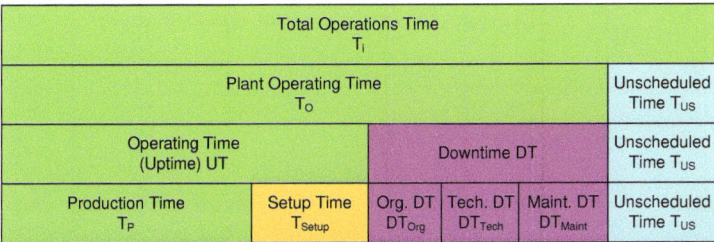

**Figure 1.7:** *Temporal Measurands*

Most production systems are not operated 24 hours a day and 7 days a week. This potential time buffer is denominated as *Unscheduled Time* $T_{US}$. When the *Downtime* $DT$ of the system is subtracted from the *Plant Operating Time* $T_O$, residual time is called *Operating Time* or *Uptime* $UT$. System downtimes can be specified into *Organisational Downtime* $DT_{Org}$, *Technical Downtime* $DT_{Tech}$, and *Maintenance Downtime* $DT_{Maint}$. $DT_{Org}$ is caused by organizational deficiencies, e.g. shortage of labor, raw material or interruption of infrastructure services. When the cause of an interruption is a technical defect, the associated downtime is the $DT_{Tech}$ (time for corrective maintenance). Time for preventive maintenance activities is assigned to maintenance downtimes $DT_{Maint}$. Usually, a production system needs some preparation time between two consecutive orders to setup and change over the production system. This preparation time is called *Setup Time* $T_C$.

**Definition 6** **Unscheduled Time** $T_{US}$ *is idle time where the system is unemployed as during weekends, company holidays, idle night shifts, etc. "The factory is closed, the lights are out".*

**Definition 7** **Setup Time** $T_{Setup}$ *is time consumed to setup production system for a new order or other product. Change-over incorporates tool or material changing, parameter adaption of production processes, etc.*

**Definition 8** *During* **Production Time** $T_P$ *the production system provides* *output regardless of the amount, speed or quality.*

A segmentation into different temporal sections allows a simple definition of another, very important term:

**Definition 9** *Utilization $U(t, T_i)$ is the ration between production system operating time and total operating time during the interval $T_i$*

$$U(t, T_i) = \frac{T_O}{T_i}$$

Utilization denotes the degree of seized capacity of a production system and is very popular in queueing theory ( [Bolch et al., 1998] and appendix B). The theory enables a mathematical description of the characteristics of a buffered system with a queue. Utilization of a queueing system should be strictly less than one to provide smooth operating. This result is directly derived from the calculation of the expected waiting time $E[WT]$ of a job in a single station queueing system [Schoensleben, 2002].

$$E[WT] \sim \frac{U(t, T_i)}{1 - U(t, T_i)} \quad (1.3)$$

If $U(t, T_i)$ approaches one, $E[WT]$ becomes infinity, queue size increases until buffer capacity is reached and blocks the precedent processes because the jobs on the preceding processes cannot leave the capacity; the manufacturing process collapse.

### 1.4.4 System and Component

Terms *system* and *component* are related in the way that a system is built up of components. The following definition is based on [VDI-3633-Blatt-1, 1993].

**Definition 10** *A **system** is a distinguishable arrangement of components building a functional entity. Components interact and provide a function which could not be performed by a single component.*

**Definition 11** *A **component** is the indivisible, atomic and smallest item in a system.*

## 1.4. Definitions and Terminology

Those definitions highlight several aspects of a system:

**Decomposability** is a crucial aspect of systems and follows a deduction, from the entirety to the nucleus. A system can be partitioned into subsystems until a non-divisible entity remains, the component. Components are regarded to be the atomic, the smallest and not further divisible items of a system. Non-divisibility may not be limited to physical laws that makes a decomposition impossible but rather incorporates economic or logistics reasons. In industrial systems, spare parts are such typical components, which could, physically, be decomposed. Decomposability gives rise to split up the system into smaller parts which are, hopefully, easier to investigate and understand. In complex systems with *emergent characteristics*, this deductive approach cannot be applied due to the fact, that when the system is divided into smaller portions, the emergent phenomenon disappears. A typical emergent phenomenon is *safety*, a property on system level and un-allocable to a single component (see [Fabricius, 2003]).

**Interaction and complexity** A system is more than the sum of its components, expressing the knowledge that the essential part of the system is its genuine composition of different components. The combination of different components gives rise to provide functionality which is inherently absent in the single components and can lead to nonlinear characteristics of the system. This extended functionality is closely related with system complexity. A system is called to be *complex* when it exhibits nonlinear and unpredictable characteristics. Those characteristics derive from two sources, large amount of system components, and its interdependencies. The difference between simple and complex system is that a complex system may expose surprising characteristics, is mostly unpredictable, expose a high degree of connectedness, is uncontrollable and decentralized to a large extend and looses main features when it gets decomposed. System complexity is unassignable to a single or a couple of components. Complexity is a system characteristics that may disappear when even a single system component is removed. Another outstanding feature of complexity is its irreducibility. By the act of reduction, emergent characteristics of the system get lost. The fast development in computer science and easy access to enormous computational power made the creation of "silicon surrogates" possible, in which the complex system can be embedded. Contemporary approaches to investigate complex systems are strongly related to modelling and simulation.

## 1.4.5 Model, Modelling and Simulation

Study of complex systems requires a virtual, abstracted representation of the system. This representation is called model.

**Definition 12** *A **model** is a schematic description of a system, theory, or phenomenon that accounts for its known or inferred properties and may be used for further study of its characteristics [Winsberg, 1999].*

A model has not the intention to be a copy of a real-world system, but to be shaped with reference to some aspects in focus. This refers to the notion of *abstraction* and highlights the fact that a model is always tailored to provide answers to a specific problem. Adequacy of a model is independent from the truth to reality but from giving the best answers and helping to understand the system. For this reason, a model is always a simplification of the reality to provide an argumentative framework that enables reasoning about the system. This simplification can even make problem investigations possible. Degree and focus of simplification is geared to the problem statement, available resources, as time, money, manpower, or computational power, and modeler's preferences. Although anything that is created with the intension to make access to a problem is considered to be a model, only models using mathematical or logic symbols to describe model characteristics are covered here.

This refers to the question of *notation*, *syntax* and *semantic* used to describe a model and to modelling languages in general. Modelling languages do not only differ from each other by different notation, syntax or semantic, but mainly from their different modelling dimensions [Schoensleben, 2001]. Dimensions highlight diverse model aspects which are:

- Procedural dimension
- Functional dimension
- Object dimension
- Task dimension

## 1.4. Definitions and Terminology

Many modelling languages combine several dimensions. However, even when a modelling language encompasses more than one dimension, there always exist a principle dimension, the primary dimension (compare with [Schoensleben, 2001]).

#### 1.4.5.1 Procedural Dimension

Process-oriented modelling describes a model with sequential processes. A process transforms the process inputs into outputs and represents a part of the system that performs an activity. The possibility of sequential arrangement of processes allows to represent temporal procedures. Referring to [DIN-66201, 1981], a process is defined as:

**Definition 13** *A **process** is the entity of interacting procedures in a system which modify or store matter, energy, and information [Meyer et al., 2006].*

Every change in the system in terms of place, time, status, or value requires a process. Temporal and logic sequences in a system can be described with state charts that represents the system dynamics.

**Definition 14** *A **state chart** is a graphical representation of the system dynamics with states and transitions. States describe stiff system conditions and the transition between the states are triggered by variations of the in- and output-flows of the associated process. When the condition of an output transition is evaluated to true, the transition fires and the action of the transition is executed [Meyer et al., 2006].*

#### 1.4.5.2 Functional Dimension

Functional modelling summarizes functions according to an chosen attribute and orders them under a superior function. The result is a tree-hierarchy with the superior function on its top being an aggregation of sub-functions. Very often, the shared attribute is having membership in the same process. The absence of time in the functional dimension

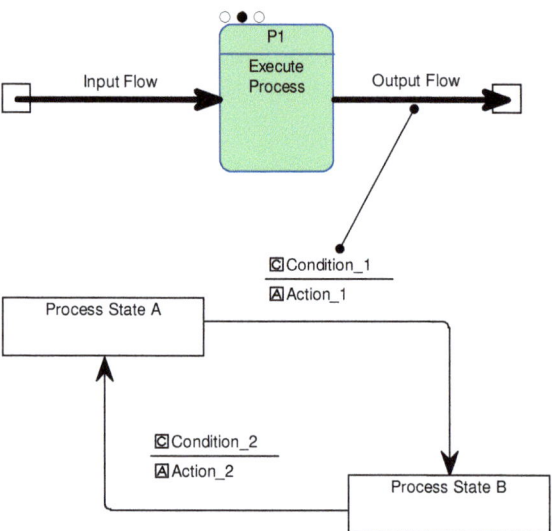

**Figure 1.8:** *Relation between Process and State Chart following [Meyer et al., 2006]*

distinguishes the procedural from the functional dimension. By contrasting function and process it can be stated that a process consists of at least two functions which are brought into a temporal sequence.

#### 1.4.5.3 Task Dimension

Task-oriented modelling originates from business organization and summarizes processes and functions that follow criteria of entirety. This bunch of tasks can be delegated to an (organisational) unit. Task dimension is strongly linked with organisational issues and is extensively used in modelling workflow management and in organization charts.

#### 1.4.5.4 Object Dimension

Object-oriented modeling abstracts data and functions, respectively processes, that concern the same *object*. The following definitions are taken from [Schoensleben, 2001]

## 1.4. Definitions and Terminology

**Definition 15** *An **object** is an instance of a class and is described with attributes and methods.*

Nearly everything can be an object as a human being, an entity, a customer, a supplier, a product, or a raw material. Attributes of an object are used to characterize and distinguish different objects of the same *class*.

**Definition 16** *An **attribute** is a characteristics, feature or a trait of an object.*

Class is an umbrella term for similar objects.

**Definition 17** *A **class** is an amount of entities or objects that share their essential characteristics and are described by the same attributes.*

Processes and functions, that belong basically only to one object, can be assigned to this specific object. Then, they turn into methods of the object.

**Definition 18** *A **method** is the ability of an object. Abilities can be processes, functions or tasks.*

A model is essentially required for *simulations*. Whereas a model can exist without simulation, it is impossible to perform a simulation without a model. Simulation is used in many contexts and multiple scientific fields in order to gain insight into the mode of operation of the model. Experiences achieved by simulation are thought to be transferable to the real world system represented by the simulation. Many definitions of the notion simulation exist (see [VDI-3633-Blatt-1, 1993], [Acel and Hrdliczka, 2002] or [Davidsson et al., 2005]). Their common ground is that a simulation is an imitation of characteristics of a real system. This representation provides knowledge about the system by varying certain parameters or aspects of the model (perform an experiment).

**Definition 19** *A **simulation** is an experiment performed on a model [Korn and Wait, 1978].*

In this context, experiment is understood as varying parameters or changing the setup of the model. Although simulation is not confined to a specific simulation engine, most simulations are run on computers. While the formal modelling of systems using mathematics tries to find analytical solutions, computer simulation can provide resolutions to problems which are beyond the limits of analytical approaches. Computer simulation uses approximation techniques (e.g. numerical integration) to find results to problems which have no analytical solution, can combine discrete and continuous system characteristics, or interacts with the simulator.

### 1.4.6 Logistics

The term logistics has its seeds in the military to supply the troops with weapons, ammunition and nutrition from their base to a forward position. Over the years, the notion has changed and found entrance to business world. Nowadays, the domain of logistics is understood to be in activities providing the customer with the right product, in the right place and at the right time. [Ghiani et al., 2004] follow the definition:

**Definition 20** *Logistics deals with the planning and control of material flows and related information in organizations, both in the public and private sectors. Broadly speaking, its mission is to get the right materials to the right place at the right time, while optimizing a given performance measure (e.g. minimizing total operating costs) and satisfying a given set of constraints (e.g. a budget constraint).*

Whereas procurement, inventory management, transportation management and so on, are important activities, logistics pursues a more holistic perspective by integrating and combining all of those activities under some restrictions (e.g. money or time). Logistics has undergone some dramatic changes in the recent years. The wide spread of JiT and accelerated product cycle urge the manufacturer to do more transactions with smaller quantities, with less lead-time, with reduced costs and with greater accuracy [Ghiani et al., 2004].

Within the framework of logistics, the terms service level $S_L(t, k)[]$ and fill rate $F_R(t, k)[]$ should be clarified. [Schoensleben, 2002] defines them as:

## 1.4. Definitions and Terminology

**Definition 21** *Service level $S_L(t,k)$ is defined as:*

$$S_L(t,k) = \frac{\text{Amount of Orders delivered in Time of } k \text{ Orders}}{k \text{ Orders}}$$
$$k = \text{Amount of Orders}$$
$$t = \text{Time}$$

**Definition 22** *Fill rate $F_R(t,k)$ is that percentage of demand that can be satisfied through available inventory or by the current production schedule. Hence, it is the number of products delivered on desired delivery date divided by the number of products ordered. A poor fill rate results in opportunity costs and, eventually, penalty costs.*

$$F_R(t,k) = \frac{\text{Amount of delivered Products}}{\text{Amount of ordered Products}}$$

Chapter 2

# Modeling of Availability Function

In manufacturing industries, two distinctive fields of application of availability engineering are differentiated: product availability/reliability and production system availability. Although they differ in perspective and their legal scope, the same assessment and analyzing techniques are applied. Product availability/reliability is strongly linked to the notion of quality and will not be in focus in the following discussion.

Modelling methods of production system availability differ in their mode of integration of temporal developments. Either, they neglect any time variance in the evolution of system availability (static approach) or they do (dynamic approach). Some static methods have been expanded to incorporate temporal effects (e.g. Dynamic Fault/Event Trees). Figure 2.1 gives a brief outlook over modelling techniques and the most applied techniques will be discussed and illustrated with examples.

## 2.1 STATIC METHODS

Static methods require less information about the system characteristics than dynamic methods. A consequence of this diminished information requirement is their application in the early phases of a project where the knowledge about the system characteristics (mainly the dynamics of the system) is very bounded. Although the knowledge base might be fairly limited, static methods provide good estimates about expected future availability. Furthermore, they are generally more understandable, intuitional, easier in utilization, and provide quicker results than their dynamic counterparts. The main drawback is their inability to deal with temporal developments. Since temporal sequences

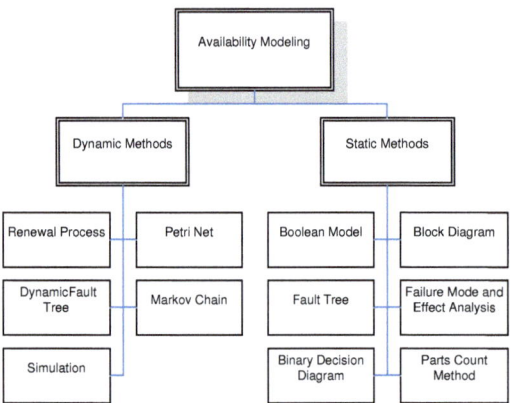

**Figure 2.1:** *Modelling Methods*

cannot be represented with static methods, they fail in modelling maintenance activities.

### 2.1.1 Boolean Model

Boolean models ( [VDI-4008-Blatt-2, 1998], [DIN-EN61078, 1994]) describe system characteristics through properties of its components and logical operations of their interplay by applying Boolean algebra. They provide binary statements about the system state in the form "in operation" and "in failure". Gradations and nuances of availability performance cannot be modelled. Following this principle of describing system characteristics brings about simplified time-independent system structures. Basis of the Boolean model is logic algebra with its Boolean variables

$$X_i, X_{i+1}, \ldots, X_n \quad \epsilon \quad [0, 1]$$

and Boolean operators

$$\begin{aligned} \text{Conjunction} &= \text{AND} \quad \wedge \\ \text{Disjunction} &= \text{OR} \quad \vee \\ \text{Negation} &= \text{NOT} \; X_i \; \overline{X_i} \end{aligned}$$

Boolean variables can only take values 1, for being "in operation", and 0, depicting "in failure". Those operators follow the rules of Boolean algebra defined in truth tables.

## 2.1. Static Methods

Truth tables are a simple (but bulky) method for showing all of the possible combinations that will lead to failure or operating state. Outcome of a truth table is a Boolean equation representing all possible failure combinations of the subsystems leading to a system failure. Those failure combinations incorporate simultaneous failures too; in the Boolean representation, coinciding failures are not neglected. Failures of the subsystems and components are expressed in terms of failure probabilities or failure rates. That means that the probability of a component being in operation is:

$$P[X_i = 1] = A_{Xi}$$
$$X_i \; : \; \text{State variable of the associated component } i$$

Unfortunately, Boolean variables in the Boolean equation cannot directly be replaced by these failure probabilities due to the residual terms of the disjunctions.

**Disjunction**
$$X_1 \vee X_2 = P[X_1] + P[X_2] - P[X_1] \cdot P[X_2] \tag{2.1}$$

**Conjunction**
$$X_1 \wedge X_2 = P[X_1] \cdot P[X_2] \tag{2.2}$$

The residual term in the disjunction is the reason for combinational term explosion. However, simplification can often be applied for sufficiently small numbers.

### 2.1.2 Block Diagram

Block Diagrams offer a graphical depiction of Boolean Equations with serial and parallel arrays of block symbols representing state variables of the underlaying subsystems. Even though a Block Diagram resembles a structural layout of a system, it is only a representation of the inherent logical coherencies without any similarities with the physical assembly of the system.

### 2.1.3 Fault Tree

A Fault Tree (FT) is a type of Boolean logical tree representing all the sequences of individual component failures that cause the system to fail in a pictorial way. Analysis starts with the definition of a single, well-defined undesirable event, which is the top-event of the tree. In case of availability studies, this unintended event is "system failure" [Roberts et al., 1981].

Starting from the top-event, FT analysis is performed in a top-down procedure by decomposing the top-event into its triggering lower level events by means of logic gates. The process of reduction stops when a basic event is reached whose probability of occurrence is below a predefined level. Basic events are assumed to be mutually independent and their probability of occurrence is described with probability distributions.

FT buildup follows a deductive, top-down approach aimed at identifying combinations of component failures that trigger top-event occurrence. Thus, a FT analysis is always tailored to a specific top-event and is not a quantitative model describing all possible events. Hierarchy levels in the FT are connected with gates (logic elements). Since these gates represent the Boolean "AND" or "OR" and the basic events can be interpreted as state variables, FTs are nothing else than graphic representations of Boolean equations in a canonical form and encounter the similar deficiencies named in 2.1.1.

Another drawback refers to the classic approach of reducing the Boolean equations into sequential combinations of events triggering a system failure. Those "successful" paths are designated to minimal cut-sets (see A.1.3.1). The number of minimal cut-sets can become enormously large in ample FTs so that many solution approaches use a cut-off frequency for cut-sets with negligible contribution. The selection of the appropriate cut-off frequency is a trade-off between result accuracy and computing time. [Nusbaumer, 2007] stated "there is no way to ensure that this approximation is accurate. For instance, if we consider a thousand cut-sets of probability $10^{-9}$ and truncate one million of cut-sets with a probability of $10^{-11}$, then we underestimate the risk by a factor $10$". Further deficiencies are:

- Limited to static representations

- Large and complex systems are difficult to survey

- Absence of accurate component failure probabilities

- Strict sequential representation of failure deployment in the system. Feedbacks and loops cannot be modelled

- No evidence that all possible failure paths are considered in the FT

- Provides no further insights about failure behavior or failure development in the system and serves only as a documentation of acquired knowledge

- Difficult to integrate advanced maintenance strategies

## 2.1. Static Methods

### 2.1.4 Binary Decision Diagram

A binary decision diagram (BDD), is a data structure that is used to represent a Boolean function or to translate a fault tree into a more efficient form for evaluation. These structures are composed from a set of binary-valued decisions, resulting in an overall decision that can be either *TRUE* or *FALSE* [Meinel and Theobald, 1998]. Each decision is predicated on the evaluation of one input variable. The data structure bases on the idea of Shannon expansion which allows splitting a switching function into two sub-functions by assigning one variable. Shannon's expansion is a method to represent a Boolean function by the sum of two sub-functions of the original. Let be $\Phi(X_1, X_2, \ldots, X_i, \ldots, X_n)$ an arbitrary Boolean function. Then, this original function can be depicted as:

$$\begin{aligned}\Phi(X_1, X_2, \ldots, X_i, \ldots, X_n) &= X_i \wedge \Phi(X_1, X_2, \ldots, \mathbf{1}, \ldots, X_n) \\ &\vee \overline{X_i} \wedge \Phi(X_1, X_2, \ldots, \mathbf{0}, \ldots, X_n)\end{aligned} \quad (2.3)$$

Shannon expansion gives rise to a new kind of normal form, the *If-then-else Normal Form (INF)*, which has a lot of useful applications. One of them is that when a value is assigned to $X_i$ only one subfunction of the expression has to be evaluated and the corresponding variable is eliminated in the remaining subfunction. For $X_i = 1$ the lower row in 2.3 disappears and the whole expression degenerates to:

$$\Phi(X_1, X_2, \ldots, X_i, \ldots, X_n) = \Phi(X_1, X_2, \ldots, 1, \ldots, X_n)$$

The variable eliminating effect of Shannon expansion offers the opportunity to finally eliminate a variable from the original function and to split up any function into subfunctions until the remaining subfunction contains only two variables. Then the eliminating process breaks up. This characteristic is used in Binary Decision Diagrams. Since the eliminated variable can be understood as "decision variable", the act of assigning a value to this variable refers to a classifying test. This test may be of the form: *Is the i-th input variable $X_i$ 0 or 1?*. In case of the result $X_i = 0$, other tests have to be proceeded than in case $X_i = 1$. If such a test or subfunction $\Phi_1$ is considered as sub-tree, it can be represented by a binary decision tree. A lot of this sub-trees are easily seen to be identical and its tempting to use their equality to reduce tree size. Then, binary decision tree is no longer a tree of Boolean expressions but can be translated into a binary decision diagram (BDD), a directed acyclic graph. Each sub-function can be interpreted as the node of a graph and is either *terminal* or *non-terminal* depending if the eliminating process terminates or not.
BDD-size is defined by the function being represented and the chosen ordering of the

variables. It is of crucial importance to care about variable ordering. Depending on the variable ordering, BDD-size may vary between a linear to an exponential range for a given function. However, the finding of the most efficient variable ordering is NP-hard [Bollig and Wegener, 1996] (NP-hard: nondeterministic polynomial-time hard problem is expressed in simple words as a problem that cannot be solved in polynomial time). BDD offers a reduction algorithm for minimizing the size of the binary decision tree but the BDD-size is very sensitive to variable ordering. Variable ordering is NP-hard, though finding an appropriate variable ordering is more an art than an engineering science. An encompassing discussion about BDD's can be found in [Nusbaumer, 2007].

## 2.2 Dynamic Methods

Strength of dynamic methods is their capability to incorporate temporal effects and developments in time. This gives rise to represent system deterioration and maintenance strategies. However, level of difficulty in use and required system information increase with their expanded applicability and tend to be less comprehensive and require more data than static approaches.

### 2.2.1 Alternating Renewal Process

A renewal process is a stochastic point process of arbitrary points of renewal on the time axis (see [VDI-4008-Blatt-8, 1984]). Differences between two subsequent renewal points are stochastic probability variables (renewal intervals) with independent identical distributed property (iid) and arbitrary probability distribution. Alternating renewal processes occur when repair or replacement times exceed an unnegligible period in time. In this case, up- and downtimes are alternating and availability is less than $1$.

Let the process begin with a new device at $t = 0$ and sequentially label the endpoints of an up-, respectively downtime, with $S_1, S_2, \ldots, S_n$ then:

$$T_{1n} = S_{2 \cdot n} - S_{2 \cdot n - 1}$$
$$n = 1, 2, \ldots, n$$

are the respective uptimes (see Figure 2.2), distributed according to

$$G_1(t) = P(T_{1n} \leq t | X_i = 1)$$
$$g_1(t) = \frac{dG_1(t)}{dt}$$

## 2.2. Dynamic Methods

The probability distribution function $G_1(t)$ represents the probability that $X_i = 1$ (compare with definition in 2.1.2).

$$T_{0n} = S_{2 \cdot n-1} - S_{2 \cdot n-2}$$
$$n = 1, 2, \ldots, n$$

the associated downtimes with the distribution

$$G_0(t) = P(T_{0n} \leq t | X_i = 0)$$
$$g_0(t) = \frac{dG_0(t)}{dt}$$

Accordingly to the definition above, the probability distribution function $G_0(t)$ stands for the probability that $X_i = 0$.

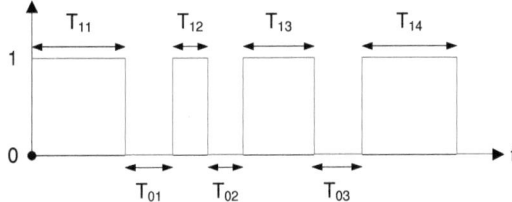

**Figure 2.2:** *Alternating Renewal Process*

Availability of an alternating renewal process is the probability $A_{Xi}(t)$ that the process is in upstate at a given point in time $t$ provided that process started in upstate at $t = 0$. In accordance to the law of total probability (see [Birolini, 2007]), $A_{Xi}(t)$ is:

$$A_{Xi}(t) = \int_0^t (g_0(h) * g_1(h)) \cdot A_X(t-h) dh + 1 - G_1(t) \quad (2.4)$$

Applying Laplace transformation on above integral equation, a closed form for $A_{Xi}(t)$ can be derived.

$$\mathcal{L}(A_{Xi}(s)) = \frac{1 - \mathcal{L}(g_1(s))}{s \cdot [1 - \mathcal{L}(g_0(s)) \cdot \mathcal{L}(g_1(s))]} \quad (2.5)$$

Evaluation of equation 2.5, requires a Laplace back-transformation into the time domain. An analytical inversion exists only for some functions. All other back-transformation make use of complex numerical approximation algorithms.

Renewal processes are not exclusively limited to model availability of single components and can be extended to multi components incorporating simultaneous failures. However, those multivariate renewal processes [Gani et al., 2003] are mathematically extremely demanding since their Laplace transformed joint availability distribution can rarely be inverted and translated into time domain [Yang and Nachlas, 2001]. Since analytical solutions for $\mathcal{L}^{-1}(A_{Xi}(s))$ can hardly be found, $\mathcal{L}^{-1}(A_{Xi}(s))$ is approximated with numerical methods. These numerical inverting approximations are very time-consuming for more than two dimensions (components).

### 2.2.2 Markov Chain Approach

A Markov chain is defined as a stochastic process with the Markov property meaning that the previous state is irrelevant to predict the probability of the sequencing state [Hermanns, 2002]. States have no causal connection with each other state and a Markov chain is "memoryless". The manner of how a certain state was reached has no influence of the future evolution of the process [Bolch et al., 1998]. In this way, a Markov chain is an extension of an alternating renewal process to more than two alternating states but limited to exponential distributed sojourn times (Semi-Markov chains provide an extension to arbitrary distributed sojourn times).

A Markov chain is a sequence of state variables

$$A, B, \ldots, N,$$

with Markov property

$$P(L = l | K = k, \ldots, A = a) = P(L = l | K = k) \qquad (2.6)$$

and switching probabilities denoted by $\lambda$ and $\mu$. Switching probabilities indicate the probabilities that the stochastic process switches from state $A_1$ to state $A_0$, respectively the other way round, during an infinite period in time. The state space of the chain is formed by all state variables.

In the field of risk and availability assessment, Markov chains offer an ease-of-use approach to model renewable systems, namely the iterating process of up- and down-times. Let Figure 2.3 bet the corresponding state diagram representation of a renewable system where $A_1$ indicates functional state and $A_0$ represents failed state of the component A.

## 2.2. Dynamic Methods

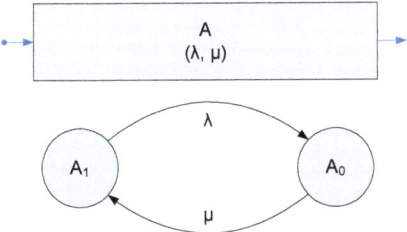

**Figure 2.3:** *State Diagram of the one Component, Renewable System*

$\lambda$ and $\mu$ are the component failure and repair rates (the switching probabilities) and are defined to be constant. State probabilities $P[A_1](t)$ and $P[A_0](t)$ can be denoted by a set of ordinary differential equations according to their temporal evolution.

$$\frac{dP[A_1](t)}{dt} = -\lambda \cdot P[A_1](t) + \mu \cdot P[A_0](t)$$
$$P[A_0](t) = 1 - P[A_1](t)$$

Application of Markov chains is difficult due to its tendency of state space explosion [Bolch et al., 1998]. Every possible state combination has to be modelled to depict system characteristics with Markov chains. Therefore, state space exhibits an exponential growth with every additional component [Hermanns, 2002].

A lot of effort has been taken to overcome state space explosion, mainly by ignoring "less important" states or neglectable state combinations [Choi et al., 2006], trying to identify subclasses or exploiting symmetries [Donatelli, 1994]. Transition times and transition probabilities are constrained to exponential distribution (constant rates) due to satisfy Markov property. However, this restriction is a severe limitation in the usability of Markov chains to model degradation processes and failure modes, since these effects are not of memory-less exponential character [Katoen, 1999]. Semi-Markov chains were established to overcome this drawback allowing any probability distribution to describe sojourn time, and transition probability, in a specific state. However, analytical tractability in Semi-Markov chains becomes very demanding and they still suffer from state space explosion.

## 2.2.3 Object Oriented Simulation

Object oriented simulation exploits the advantages provided by object oriented programming over procedural programming. This advantage is mainly the efficiency in writing a simulation [Joines and Roberts, 1998]. A code written in an object oriented style is fairly similar to the same procedural code but brocken down into several smaller pieces as opposed to a traditional view in which a program may be seen as a group of tasks ("subroutines"). The idea behind object oriented programming is a careful application of abstractions and subdividing problems (top-down approach) into smaller and less complex subtasks that are assigned to "objects". This hierarchical approach gives rise to the possibility of individually modelling inherent object characteristics (mostly by state charts) independently from the interplay between those objects on the same hierarchy. Though, interaction between different objects has only to be represented with respect to the associated hierarchical level, and mode of interaction can be subdivided into smaller behavioral patterns. Impact from higher level object characteristics on lower hierarchical levels are automatically integrated by inheritance. This simplifies modelling in the way that possible "state combinations" arising from higher level states don't have to be modelled explicitly but emerge.

Unlike in procedural programming, object oriented programming starts with determining all participating elements (these are the objects in the model). Next step is the modelling of the dynamics and characteristics of the objects and its modality of interaction with respect to their hierarchical level. This procedure is redone until the model is adequately detailed.

Procedural programming refers more to long sequences of statements and those instructions generally have a direct correspondence to a form of flowchart (sequential processing of instructions). Any hierarchical decomposition approach is absent (no top-down methodology). Missing hierarchy is the major source of the eventually greatest limitation of the procedural style: extensibility. The only way to adapt and extend those simulations is through functional extension. Structural functionality can be added but any changes of the design of the simulation causes a complete reprogramming since already existing structures can hardly be reused.

The characteristic of inheritance limits the amount of simultaneous failures that have to be modelled on a specific level of hierarchy. Within a certain level, only the failure combinations arising on that level, have to be modelled explicitly. All other combinations on higher level emerge due to inheritance and can be neglected. Object oriented simulation offers the possibility to simulate very complex systems as e.g. the model presented in chapter 5.

## Chapter 3

# Connection between System Failure and Logistics

The following chapter is dedicated to investigate and explain the interrelations between system failure (system availability) and its impact on logistic objectives. So far, this is the first pure analytic approach to represent the connection between system failure and logistics.

## 3.1 MATCHING OF OUTPUT AND DEMAND

Principle task of logistics is to provide customers with products in sufficient quantity and in due time. An adequate matching of production output $O(t, T_i)$, inventory $I(t, T_i)$ and demand $D(t, T_i)$ asks for planning and forecasting both, production and demand. A short introduction into demand forecasting techniques can be found in appendix C. Demand, production and inventory have to be in balance to ensure that all demand can be satisfied:

$$D(t, T_i) \leq O(t, T_i) + I(t, T_i) \qquad (3.1)$$

with

$D(t, T_i)$ : Demand function in $t$ during interval $T_i$ [pc.]
$O(t, T_i)$ : Production function in $t$ during interval $T_i$ [pc.]
$I(t, T_i)$ : Inventory function in $t$ during interval $T_i$[pc.]
$I(t, T_i) \geq 0 \ \ t \ \epsilon \ [0, \infty]$

Using equation 3.1, non-deliverability can be defined as:

$$D(t, T_i) > O(t, T_i) + I(t, T_i) \tag{3.2}$$

Within this section, all parameters in equation 3.1 get analytically modelled to show the interrelationship between those factors. Furthermore, it will be shown how logistic requirements set additional constraints to system availability.

### 3.1.1 Production Output $O(t, T_i)$

Production output $O(t, T_i)$ during period $T_i$ can be written as:

$$O(t, T_i) = U(t, T_i) \cdot A_{Mission_S}(t, k, T_i) \cdot \frac{T_i}{CT} \tag{3.3}$$

with

$U(t, T_i)$ : Utilization of the production system during $T_i$

$A_{Mission_S}(t, k, T_i)[]$ : Average system mission availability during $T_i$

$$\approx \prod_{l=1}^{k} A_{Mission_l}(t, T_i)$$

$A_{Mission_l}(t, T_i)$ : Random variable of mission availability during $T_i$ of component $l$

$$= \frac{T_i - DT_l(t, T_i)}{T_i}$$

$T_i[h]$ : Time interval

$DT_l(t, T_i)[h]$ : Cumulative downtime of component $l$

$CT$ : Cycle Time $\left[\frac{h}{pc.}\right]$

Let be $U(t, T_i) = 1$, then equation 3.3 is:

$$O(t, T_i) = A_{Mission_S}(t, k, T_i) \cdot \frac{T_i}{CT} \tag{3.4}$$

It is assumed, that $CT$ is fix but $A_{Mission_S}(t, k, T_i)$ may vary.

#### 3.1.1.1 Analytical Modeling of $A_{Mission_S}(t, k, T_i)$

Mission availability of the system is defined as the fraction of period $T_i$ in which the production system is in operation.

$$A_{Mission_S}(t, k, T_i) = \frac{T_i - DT_{System}(t, T_i)}{T_i} \tag{3.5}$$

## 3.1. Matching of Output and Demand

with

$DT_{System}(t, T_i)[h]$ : Cumulative system downtime during $T_i$

On this account, downtimes of an interlinked system are modeled analytically, and resulting equation is used to compute mission availability $A_{Missions}(t, k, T_i)$. It is assumed that the system and the single component characteristics can simplified be described with two states, *Uptime* and *Downtime*, and that those components are in series.

If the single components form a serial system are independent and rare-event approximation is applied, system downtime $DT_{System}(t, T_i)[h]$ can be written as the sum of component downtimes $DT_l(t, T_i)[h]$ (compare with [Lim et al., 2005]):

$$\begin{aligned} DT_{System}(t, T_i) &\approx DT_1(t, T_i) + DT_2(t, T_i) + \ldots + DT_k(t, T_i) \\ &\approx \sum_{l=1}^{k} DT_l(t, T_i) \end{aligned} \quad (3.6)$$

with

$DT_l(t, T_i)[h]$ : Cumulative downtime of component $l$ during $T_i$

Since all other assembling layouts can be replaced by an equivalent serial element (compare with section A.3.1), equation 3.6 can be applied as approximation for any other subsystem type.

Let be

$MTTR_l(t, T_i)[h]$ : Mean time to repair of component $l$ during $T_i$
$F_l(t)[h]$ : Uptime distribution of component $l$
$G_l(t)[h]$ : Downtime distribution of component $l$
$N_{l0}(t, T_i)[]$ : Amount of interruptions of component $l$ during $T_i$

In particular, the expected cumulative downtime of a single component $DT_l(t, T_i)$ is the combination of two random events, the amount of interruptions $N_{l0}(t, T_i)$ and the expected $MTTR_l(t, T_i)$ during period $T_i$ (see renewal process in 2.2.1). These items can be modeled separately. Let $F_l(t)$ and $G_l(t)$ form an alternating renewal process then $N_{l0}(t, T_i)$ is ( [VDI-4008-Blatt-8, 1984] in combination with [Birolini, 2007]):

$$N_{l0}(t, T_i) = \int_{T_i} \sum_{n=1}^{\infty} F_l * (F_l * G_l)^{n-1}(t) dt \quad (3.7)$$

and

$$MTTR_l(t, T_i) = \int_{T_i} (1 - G_l(t))\, dt \qquad (3.8)$$

Then, $DT_l(t, T_i)$ can be written as:

$$DT_l(t, T_i) = N_{l0}(t, T_i) \cdot MTTR_l(t, T_i)$$
$$DT_l(t, T_i) = \int_{T_i} \left( \sum_{n=1}^{\infty} F_l * (F_l * G_l)^{n-1} \right) \cdot (1 - G_l(t))(t)\, dt \qquad (3.9)$$
$$\text{for } n \to \infty$$

This equation 3.9 can be inserted in equation 3.6:

$$DT_{System}(t, T_i) = \sum_{l=1}^{k} \int_{T_i} \left( \sum_{n=1}^{\infty} F_l * (F_l * G_l)^{n-1} \right) \cdot (1 - G_l(t))(t)\, dt \qquad (3.10)$$
$$\text{for } n \to \infty$$

Probability variable $A_{Missions}(t, k, T_i)$ can be achieved by combining equation 3.5 with equation 3.10.

$$A_{Missions}(t, k, T_i) = \frac{T_i - \sum_{l=1}^{k} \int_{T_i} \left( \sum_{n=1}^{\infty} F_l * (F_l * G_l)^{n-1} \right) \cdot (1 - G_l(t))(t)\, dt}{T_i}$$
$$\text{for } n \to \infty \qquad (3.11)$$

### 3.1.2 Inventory

A stock acts as a buffer between production and demand and may balance temporal or quantitative variations in order to provide a high service level and fill rate. Thus, inventory synchronizes production and demand. Whenever the lead time claimed by the customer is lower than the actual lead time, stockkeeping is necessary (see Figure 3.1). This lead time claimed by the customer defines the order entry point, the production concept and the stocking level. The stocking level is that level in the value-added chain above which a product can be produced within the delivery lead time required by the customer. For items below and at the stocking level, no exact demand is known and demand forecast is necessary (see section C).
According to [Schoensleben, 2002], four distinctive production concepts can be distinguished:

## 3.1. Matching of Output and Demand

**Engineer-to-order** No stockkeeping, development of the product or some parts of the product starts when an order is placed (e.g. plant engineering, building industry).

**Make-to-order** Engineering and design process of the product are completed but customer may select from a vast variety of options. Either raw materials are stocked or the material is directly purchased after receipt of a customer order. Stocking is at level of finished product and process development (e.g. automotive or aviation industry).

**Assemble-to-order** Stockkeeping on level of assemblies or single parts. The product is assembled according to the customer order. Customer may have a selection of some additional options to chose from (e.g. insurance industry, electrical industry, energy industry).

**Make-to-stock** means stockkeeping at the level of end products. Incoming demand will not affect production process but is satisfied from the inventory (pharmaceutical industry, food industry, most sectors in the consumer industry).

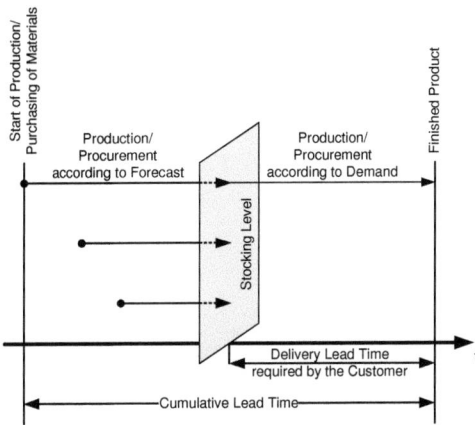

**Figure 3.1:** *Stocking Level according to [Schoensleben, 2002]*

Inventory management is a key issue in logistics and promotes the following benefits [Ghiani et al., 2004]:

**Increasing service level** Stockkeeping may balance uncertainties in demand forecast and shortens lead time. Therefore, inventory is a means to cope with randomness in customer demand and lead times.

**Reducing logistics cost** Freight transportation is dominated by economies of scale. It might be inefficient and too costly to transport small orders over long distance with high frequency. Thus, small orders get pooled to make use of economy of scale.

**Seasonality of products** Seasonal products can be held at stock to make them available when a huge demand occurs in a very short period in time which cannot be satisfied with the actual production rate (e.g. demand on chocolate Easter bunnies).

**Speculating on prices** In markets with high variability, it may be economic to purchase materials in advance when the prices are low.

**Tackling management weaknesses** Stockkeeping can be used to cover managerial deficiencies in e.g. demand forecasting and coordinating demand and supply.

State-of-the-art stockkeeping and inventory management is aimed at balancing service level, associated costs of being non-deliverable, and warehousing costs.

Inventory depends on demand $D(t, T_i)$, current quantity of production $O(t, T_i)$ and inventory $d$ at the beginning of the observation in the way that:

$$I(t, T_i) = O(t, T_i) - D(t, T_i) + d \qquad (3.12)$$

with

$$d \;:\; \text{Inventory at } t = 0$$

Insert equation 3.12 into equation 3.1:

$$D(t, T_i) \leq (2 \cdot O(t, T_i) - D(t, T_i) + d)$$

Equation 3.13 offers a representation of equation 3.1 with $D(t, T_i)$, $O(t, T_i)$, and $d$ (see subsection 3.4).

## 3.1.3 Demand

Aggregated customer demand over a time interval $T_i$ is the sum of several single events (customer orders) within this time-frame. Those single events can be described by a frequency distribution of the single events and by the distribution of its characteristic values (demand quantity). Demand is the product of the expected value of the frequency distribution of these events $P(t, T_i)$ and the expected value of the distribution of order quantities $Q(t, T_i)$ (compare with [Ferschl, 1964] and [Axsaeter, 2006]). Let be

$P(t, T_i)$ : Probability distribution of amount of orders $\left[\frac{1}{h}\right]$

$Q(t, T_i)$ : Probability distribution of order quantity $[pc.]$

$E[P(t, T_i)]$ : Expected value of the probability distribution of amount of orders $\left[\frac{1}{h}\right]$

$E[Q(t, T_i)]$ : Expected value of the probability distribution of order quantity $[pc.]$

Resulting demand $D(t, T)$ is

$$D(t, T_i) = E[P(t, T_i)] \cdot E[Q(t, T_i)] \tag{3.13}$$

$CV[P^{D(t,T_i)}]$ corresponds to the coefficient of variation of the distribution $P^{D(t,T_i)}$, i.e., the quotient of standard deviation and expected value.

$$CV[P^{D(t,T_i)}] = \sqrt{\frac{1 + CV^2[Q(t, T_i)]}{CV^2[P(t, T_i)]}} \tag{3.14}$$

Regular demand is a basic prerequisite for reliable production planning and for simple control techniques such as Kanban [Halevi, 2001]. Furthermore, most demand forecasting techniques expect normal distributed demand (regular demand) to provide reliable forecasts. A normal distribution can be assumed when the coefficient of variation of the consumption distribution $CV[P^{D(t,T_i)}]$ is equal or lower than 0.4 [Schoensleben, 2002].

$$CV[P^{D(t,T_i)}] \leq 0.4 \tag{3.15}$$

Since time $t$ is out of control, the only parameter to impact $CV[P^{D(t,T_i)}]$ is the time interval $T_i$. This time interval $T_i$ has to be chosen in accordance with equation 3.15 that a regular demand can be assumed. If the time interval $T_i$ is chosen to short, this quickly results in lumpy and uneven demand patterns. Thus, requirement of smoothed demand sets a lower boundary for planning purposes.

## 3.2 Impact of System Failure on Service Level

In logistics, deliverability is monitored by a controlling parameter called service level $S_L(t,k)$. The service level, or level of service, is the percentage of order cycles $k$ that the company will go through without stockout, meaning that inventory is sufficient to cover demand. In the case of no stockkeeping, the service level defines the percentage of order cycles in which demand can be satisfied with the production output.

$$S_L(t,k) = 1 - \frac{1}{k} \cdot \sum_{j=1}^{k} P(\text{Delayed Order } j) \qquad (3.16)$$

where

$P(\text{Delayed Order } j)$ : Probability of a delay of order $j$
$k$ : Amount of order cycles

An order $j$ is delayed when $DT_{System}(t, T_i)$ is larger than the disposal time $sT_j$ in the respective period in time $T_i$. Let be $LT_j = T_i$, then:

$$DT_{System}(t, LT_j) > cLT_j \cdot (1 + s - U(t, cLT_j)) \qquad (3.17)$$

with

$LT_j[h]$ : Lead time for order $j$ (actual lead time)
$cLT_j[h]$ : Calculatory lead time for order $j$
$s[\,]$ : Safety factor $\in [0, \infty]$
$U(t, cLT_j)[\,]$ : Utilization in $t$ during interval $cLT_j$

$cLT_j$ can be approximated with:

$$cLT_j = A_{Missions}(t, k, cLT_j) \cdot LS_j \cdot CT \qquad (3.18)$$

with

$LS_j$ : Lot size of order $j$

Available safety time $sT_j$ is a combination of unscheduled time due to an utilization below 1 (see definition 1.4.3) and a temporal safety margin $s \cdot cLT_j$. Safety lead time $sT_j$ for order $j$ is:

$$\begin{aligned} sT_j &= cLT_j \cdot (s + (1 - U(t, cLT_j))) \\ &= cLT_j \cdot (1 + s - U(t, cLT_j)) \end{aligned} \qquad (3.19)$$

## 3.2. Impact of System Failure on Service Level

### 3.2.1 Equation Derivation

Probability density of system downtime $f_{DT_{System}}(t)$ of $DT_{System}(t, LT_j)$ is:

$$f_{DT_{System}}(t) = \frac{DT_{System}(t, LT_j)}{\int_{LT_j} DT_{System}(t, LT_j)} \quad (3.20)$$

The probability that $DT_{System}(t, LT_j) > sT_j$ appears can be derived from equation 3.20:

$$P(DT_{System}(t, LT_j) > sT_j) = 1 - P(DT_{System}(t, LT_j) \leq sT_j)$$
$$= 1 - \int_0^{sT_j} f_{DT_{System}}(t) dt \quad (3.21)$$

Then, the probability of a delayed delivery of the first-ever order 1 is:

$$P(\text{Delayed Order 1}) = 1 - P(DT_{System}(t, LT_1) \leq cLT_1 \cdot (1 + s - U(t, cLT_1)))$$
$$= 1 - \int_0^{cLT_1 \cdot (1+s-U(t,cLT_1))} f_{DT_{System}}(t) dt \quad (3.22)$$

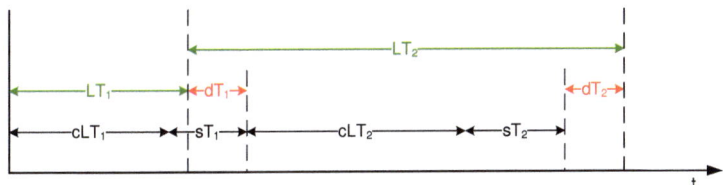

**Figure 3.2:** *Integral Intervals*

Probability of a delayed second delivery is similarly derived:

$$P(\text{Delayed Order 2}) = 1 - P(DT_{System}(t, LT_2) \leq cLT_2 \cdot (1 + s - U(t, cLT_2)) + dT_1)$$
$$= 1 - \int_0^{cLT_2 \cdot (1+s-U(t,cLT_2))+dT_1} f_{DT_{System}}(t) dt$$
$$dT_1 = cLT_1 + sT_1 - LT_1$$
$$= cLT_1 \cdot (1 + s - U(t, cLT_1)) - LT_1 \quad (3.23)$$

$dT_1[h]$ represents the additional spare unscheduled time of the previous production order which is placed at disposal of the subsequent order. $dT_{k-1}$ is:

$$dT_{k-1} = \sum_{j=1}^{k-1} cLT_j \cdot (1 + s - U(t, cLT_j)) - LT_j \quad (3.24)$$

The probability of delay of the $k^{th}$-order can be calculated by combining equations 3.23 and 3.24:

$$P(\text{Delayed Order } k) = 1 - \int_0^{a(k)} f_{DT_{System}}(t) dt \qquad (3.25)$$

$$a(k) = \sum_{j=1}^{k} cLT_j \cdot (1 + s + U(t, cLT_j)) - \sum_{j=1}^{k-1} LT_j$$

Equation 3.25 is inserted into 3.16 to calculate service level:

$$S_L(t, k) = \frac{1}{k} \cdot \sum_{j=1}^{k} \left( \int_0^{a(k)} f_{DT_{System}}(t) dt \right) \qquad (3.26)$$

Service level $S_L(t, k)$ soars with increasing safety factor $s$ and decreases with rising amount of orders $k$. High utilization and an insufficient safety factor cause that the service level approaches zero with increasing amount of orders.

Equation 3.18 can be inserted into equation 3.26 to show the dependency between $A_{Missions}(t, k, cLT_j)$ and $S_L(t, k)$.

## 3.3 Safety Factor $s$ and its Lower Boundary

Safety factor $s$ is used to mitigate and balance uncertainties in the system in order to achieve a certain level of service level. In principle, those uncertainties derive from the variability of demand $D(t, T_i)$, the fluctuations of the mission availability $A_{Missions}(t, k, T_i)$ and the impreciseness of the estimate of their mean values and variance (see equation 3.1 or 3.13).

### 3.3.1 Minimal Safety Factor $s_{min}$

Equation 3.25 approaches 1 if the integral disappears or the function $f_{DT_{System}}(t)$ is equal to 0. Integral is zero for $a(k) = 0$. Thus, for the lower boundary of $s_{min}$ holds:

$$\sum_{j=1}^{k} cLT_j \cdot (1 + s_{min} + U(t, cLT_j)) - \sum_{j=1}^{k-1} LT_j \geq 0$$

## 3.3. Safety Factor $s$ and its Lower Boundary

This equation can be solved for $s_{min}$:

$$s_{min} \geq \frac{\sum_{j=1}^{k} cLT_j \cdot (U(t, cLT_j) - 1) + \sum_{j=1}^{k-1} LT_j}{\sum_{j=1}^{k} cLT_j} \quad (3.27)$$

Any assigned value equal or below this critical measurement will immediately cause $S_L(t, k) = 0$. $cLT_j$ in equation 3.27 can be replaced with $A_{Missions}(t, k, cLT_j) \cdot LS_j \cdot CT$ to express $s_{min}$ in terms of system mission availabilities.

$$s_{min} \geq \frac{\sum_{j=1}^{k} A_{Missions}(t, k, cLT_j) \cdot LS_j \cdot CT \cdot (U(t, cLT_j) - 1) + \sum_{j=1}^{k-1} LT_j}{\sum_{j=1}^{k} A_{Missions}(t, k, cLT_j) \cdot LS_j \cdot CT} \quad (3.28)$$

Equation 3.28 is evaluated with the values recorded during the 01.03.2006 till 28.02.2007 at the industry partner. See results in table 3.3.1.

It can be concluded that:

- A minimum lower boundary for the safety factor $s$ exists. Any assigned values below this lower boundary has a negative impact on the service level.
- If the safety factor $s$ is chosen too low, the service level $S_L(t, k)$ approach 0 with increasing amount of orders $k$. The same effect has a production system utilization $U(t, T_i)$ close to 1. A utilization of 1 avoids any idle time of the production

|           | $s_{min}$ |
|-----------|-----------|
| March     | 0.93      |
| April     | 1.00      |
| May       | 0.95      |
| June      | 0.82      |
| July      | 0.84      |
| August    | 0.85      |
| September | 0.93      |
| October   | 0.77      |
| November  | 0.85      |
| December  | 0.78      |
| January   | 0.76      |
| February  | 0.95      |

**Table 3.1:** *Minimum Safety Factor $s_{min}$ at Huba Control AG*

system between two consecutive production orders and minimizes the available idle time for downtimes.

- Convolution in $f_{DT_{System}}(t)$ and the sum of the upper integral boundary in 3.26 link system failures in the past with present order dispatch.

- The analytical approach to link the service level with system downtimes (see equation 3.26) of a production system is very limited. Infinite convolution causes the need of a Laplace transformation. The back- transformation into the time domain and the calculation of the sum of integrals are very time-consuming.

## 3.4 Context of Forecasting Demand and Mission Availability

Whenever the delivery lead time required by the customer is at least as long as the cumulative lead time, products can be engineered, purchased, or manufactured at the occurrence of demand. In this case, neither stockkeeping nor demand forecasting is necessary; demand $D(t, T_i)$ is called to be purely deterministic. However, in any other situation, goods, as semifinished parts, single parts, assemblies, and raw materials, must be purchased before an order is placed. Those items have to be procured and stocked on the basis of demand forecast. The level, on which the goods are held at stock, is called stocking level (see Figure 3.1). Stocking level coincides with the order entry point, the point in time in the value-added chain when stochastic demand turns into an order.

Equation 3.13 links demand $D(t, T_i)$ and production output $O(t, T_i)$. This connection requires a planning of demand and production output. Since deviation of production output is mainly impacted by the temporal variability of $A_{Missions}(t, k, T_i)$ and the utilization $U(t, T_i)$, production output planning reduces to manage $A_{Missions}(t, k, T_i)$ and adapt $U(t, T_i)$.

Those two parameters affect production output in different time scales. Whereas utilization $U(t, T_i)$ can instantaneously be varied, $A_{Missions}(t, k, T_i)$ is more reluctant and less reactive. The expected mean value of $A_{Missions}(t, k, T_i)$ is strongly linked to the change of the system's failure rate, which exposes a very slow evolution. Modulation of the mean value requires a modification of the production system or an adaption of the maintenance strategy which needs some time to take effect. This temporal reluctance between adaption of the maintenance strategy and its impact on mission availability make an efficient use of different maintenance strategies to balance demand and production output in short term impossible. This short term demand variability can only

## 3.4. Context of Forecasting Demand and Mission Availability

be managed by stockkeeping or utilization adaption and derive from the variance of the demand distribution, mainly. In the case of no stocking, utilization modulation is used to compensate the different variances of demand and mission availability distribution. Temporal variation in mission availability can only partially, or even not at all, be compensated by an increase in utilization if utilization is close to 1. In such a situation, a decrease in variance of the mission availability is necessary to provide high service level and fill rate. An adequate matching of demand and mission availability asserts a claim of following a normal distribution for both, demand and mission availability distribution of the system.

Equation 3.13 indicates that a matching between demand and production output is required. A balancing between demand and production is needed in any kind of logistics but the available time frame may differ. In the case of stockkeeping, this time frame is usually larger than in a Just-In-Time logistics since the stock acts as a buffer.

$$D(t,T_i) \leq 2 \cdot O(t,T_i) + d$$

Linking between demand and production results in the relation that the output distribution $P^{O(t,T_i)}$ follows

$$CV[P^{D(t,T_i)}] \approx CV[P^{2 \cdot O(t,T_i)}] \qquad (3.29)$$

since

$$CV\left[\frac{P^{2 \cdot O(t,T_i)}}{s}\right] = CV[P^{O(t,T_i)}]$$

with

$$P^{D(t,T_i)} \;:\; \text{Demand distribution}$$
$$P^{O(t,T_i)} \;:\; \text{Output distribution}$$

In combination with the claim of regular demand (see equation 3.15), $CV[P^{2 \cdot O(t,T_i)}]$ must follow:

$$CV[P^{2 \cdot O(t,T_i)}] \leq 0.4 \qquad (3.30)$$

Regarding equation 3.4, temporal characteristics of the output is solely influenced by $A_{Missions}(t,k,T_i)$. If $O(t,T_i)$ in equation 3.15 is replaced with equation 3.4, then the distribution $P^{2 \cdot A_{Missions}(t,k,T_i)}$ of the random variable $A_{Missions}(t,k,T_i)$ must obey:

$$CV[P^{2 \cdot A_{Missions}(t,k,T_i)}] \leq 0.4 \qquad (3.31)$$

### 3.4.1 Planning Horizon $T_{PH}$

Any demand forecasting technique is only applicable in the case of regular demand whereas regular demand is characterized by exposing any regularities and following a normal distribution (see C). In the case of too short time interval, demand can quickly be discontinuous. This can be overcome by stretching the time span $T_i$. Since production output and demand should be in balance, requirements for assuming regular demand are applicable also for production determination. A lower minimal time interval $T_{PH}[h]$ (called planning horizon) has to be chosen in accordance to derive normal probability distributions for both, mission availability of the system and demand distribution in this period. Thus, mission availability distribution sets an additional constraint to determine minimal $T_{PH}$.

$$T_{PH} \ni T_i \begin{cases} CV[P^{2 \cdot A_{Mission_S}(t,T_i)}] \leq 0.4 \\ CV[P^{D(t,T_i)}] \leq 0.4 \end{cases}$$

Chapter 4

# Classic Maintenance Models and Failure Rate Shape

In this chapter, classic approaches to model preventive maintenance and its impact on the temporal progress of the failure rate are discussed. Most maintenance models make use of modelling techniques which are capable to represent at least two states; an operating and a failure state (see chapter 2) with transition rates between those two states. This offers the possibility to model the impact of preventive maintenance by an adaption of the transition rate between the failure and the operating state.
The second part of the chapter discusses the influence of preventive maintenance on system availability for different failure rate shapes; increasing, constant and decreasing failure rate.

**Definition 23** *Failure mechanisms are associated with the source of failures as wear-out, ageing, fatigue, or corrosion, and are related to physical, chemical or other processes that lead to failure.*

If a certain level of degradation is reached, a system failure occurs.

**Definition 24** *Failures are the result of failure mechanisms. After a failure occurs a maintenance task has to be performed to reset production system into operation.*

The way in which the failure occurs is defined as failure mode.

**Definition 25** *Failure Modes describe the final state of the degradation process and represent the way in which the failure happens and its impact on plant operation. Failure modes have to be determined in relation to the preset performance standards of the system description. For example, undue oscillation, does not open/ close, remains open/ close, unintended task execution, does not start/ stop, or shortcuts are failure modes of an arbitrary system.*

## 4.1 CLASSIC PREVENTIVE MAINTENANCE MODELS

Preventive maintenance summarizes all maintenance activities which are not triggered by a system failure. Not only the mode of maintenance task (preventive or corrective maintenance) and it is associated maintenance interval impacts the failure rate but also its level of quality (effectiveness of maintenance task). The state after a maintenance action was performed on a component is assumed to be: *perfect, imperfect, minimal, worse* or *worst* [El-Ferik and Ben-Daya, 2006]:

| Perfect Maintenance | The system state is restored to be "as good as new" | Decreasing of the failure rate |
|---|---|---|
| Imperfect Maintenance | A maintenance action that restores the system to a state somewhere between "as good as new" and "as bad as old" | Decreasing of the failure rate |
| Minimal Maintenance | The system state is "as bad as old" | No effect on the failure rate |
| Worse Maintenance | System is in operating state worse than just prior to the maintenance action | Increasing of the failure rate |
| Worst Maintenance | System breaks down right after maintenance action | Increasing of the failure rate |

Quality levels worse and worst maintenance are related to maintenance and repair induced failures and will be discussed in section 5.1.2.2. Preventive maintenance actions, as cleaning or greasing, mitigate the deterioration effect of some failure mechanics

## 4.1. Classic Preventive Maintenance Models

and restore components to a "as good as new" condition with respect to some failure mechanisms only. All other failure mechanisms will remain unaffected. Hence, Lin et. al. [Lin et al., 2001] introduced the concept of two categories of failure mechanisms, *maintainable failure mechanisms* and *non-maintainable failure mechanisms*. Preventive maintenance will effect maintainable failure mechanisms exclusively, whereas non-maintainable failure mechanisms remain unaltered.

Zequeira [Zequeira and Bérenguer, 2005] stated that the maintainable and non- maintainable failure rates are dependent. Non-maintainable failures are supposed to provoke maintainable failures and contrariwise which implies that a coupling mechanism exists. It is reasonable to believe in such a mechanism; wearout, for example, may remove protective coating of an element, promotes oxidation or corrosion that weakens the structure and could cause a fraction. Whereas wearout can be affected by preventive maintenance, no maintenance activity exists to protect a structure from fraction. Such a mechanism induces a non-maintainable failure that could never be observed if this coupling mechanism is absent. The system exposes unintended failure mechanisms. Existence of such unintended failure mechanisms would provide a strong evidence of such a coupling mechanism. Regarding the diminutive amount of observed unintended failures in the partner company, it can be reasoned that there are good arguments to believe in the existence of a coupling mechanism but that its impact on the failure rate is negligible.

Let assume that the failure rates related to maintainable and non-maintainable failure mechanisms are independent. Then, the system failure rate $\lambda_{Failure}(t)$ is:

$$\lambda_{Failure}(t) = \lambda_{Maint}(t) + \lambda_{Non-maint}(t) \tag{4.1}$$

In the literature, three approaches for modelling the impact of preventive maintenance on the failure rate have been studied extensively; a failure rate model by Lie and Chun [Lie and Chun, 1986] and Nakagawa [Nagakawa, 1986], [Nagakawa, 1988], an age reduction model by Canfield [Canfield, 1986] and Malik [Malik, 1979] and a hybrid model by Lin. et. al. [Lin et al., 2001].

### 4.1.1 Failure Rate Preventive Maintenance Model

The failure rate function after the $i^{th}$ preventive maintenance action is $\alpha_i \lambda_{i-1}(t)$ for $t \in (0, t_{i+1} - t_i)$ when it was $\lambda_{i-1}(t)$ for $t \in (0, t_i - t_{i-1})$ where $\alpha_i$ is the adjustment factor due to the $i^{th}$ preventive maintenance action. Each preventive maintenance resets the

# Chapter 4. Classic Maintenance Models and Failure Rate Shape

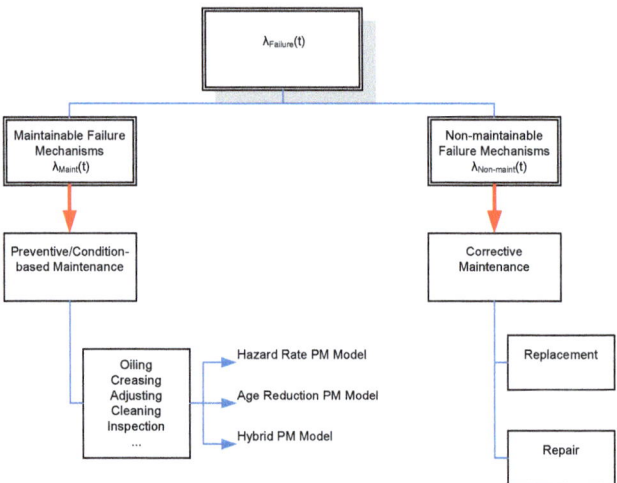

**Figure 4.1:** *Categories of Failure Mechanisms*

failure rate to the start value $\lambda_0(t)$ and the rate of increase $\alpha_i$ rises after each preventive maintenance task. The adjustment factor $\alpha_i$ is an index for measuring the quality of the preventive maintenance task and increases after every preventive maintenance.

$$\begin{aligned}
\lambda_i(t) &= \alpha_i \cdot \lambda_{i-1}(t) \\
&= \prod_{k=1}^{i} \alpha_k \cdot \lambda_0(t) \\
\lambda_0(t) &= m \cdot t \\
m &= \frac{\triangle \lambda_0(t)}{\triangle t} \\
\alpha_{i-1} &< \alpha_i \text{ with } \alpha_0 \leq 1
\end{aligned}$$

## 4.1.2 Age Reduction Preventive Maintenance Model

The concept of effective age models the impact of a preventive maintenance task in the way that the health condition of a component with a calender age of $E_i$ years is "as good

## 4.1. Classic Preventive Maintenance Models

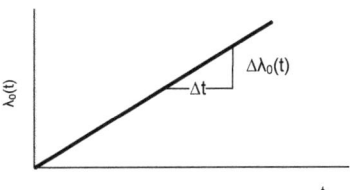

**Figure 4.2:** *Definition of $\lambda_0(t)$*

as a $\beta_i \cdot E_i$ year old" [Kijima and Suzuki, 1988], [Kijima, 1989]. Preventive maintenance makes the system younger. In terms of mathematics, the $i^{th}$ preventive maintenance task reduces the effective age $E_i$ to $\beta_i \cdot E_i$ with $0 \leq \beta_i < 1$. In this model, the value of $\beta_i$ is an indicator of the achieved quality level of the preventive maintenance activity.

$$\lambda_i(t) = \lambda_{i-1}(\beta_i \cdot t_i + t)$$
$$\text{for } t \in (0, t_{i+1} - t_i)$$

### 4.1.3 Hybrid Model

Lin, Zuo, and Yam [Lin et al., 2001] propose a hybrid preventive maintenance model incorporating the advantages of the age reduction and failure rate preventive maintenance model. Hybrid model assumes that the effects of a preventive maintenance task are modelled by two aspects:

- Long-term effect when the component is set into operation again ($\alpha_i$)
- Immediate impact after the preventive maintenance is accomplished ($\beta_i$)

The failure rate after the $i^{th}$ preventive maintenance activity becomes $\alpha_i \cdot \lambda_{t_{i-1}}(\beta_i t_i + t)$, where $t_i$ is the time when the $i^{th}$ preventive maintenance is performed and $t \in (0, t_{i+1} - t_i)$. $1 = \alpha_0 \leq \alpha_1 \leq \ldots \leq \alpha_i \leq \ldots \leq \alpha_{N-1}$ and $0 = \beta_0 \geq \beta_1 \geq \ldots \geq \beta_i \ldots \geq \beta_{N-1} < 1, t > 0$ and $\lambda_0(t)$ is the failure rate of the system without preventive maintenance (compare with figure 4.3). The parameters $\alpha_i$ and $\beta_i$ play the same role as in the age reduction and failure rate preventive maintenance model.

$$\lambda_{t_i}(t_i + t) = \alpha_i \lambda_{t_{i-1}}(\beta_i t_i + t)$$
$$\text{for } t \in (0, t_{i+1} - t_i)$$

Above equation from [Lin et al., 2001] can be expanded to:

$$\lambda_1(t_1 + t) = \alpha_1 \cdot \lambda_0(\beta_1 \cdot t_1 + t)$$
$$\lambda_1(t_2) = \alpha_1 \cdot \lambda_0(\beta_1 \cdot t_2)$$
$$\lambda_2(t_2 + t) = \alpha_2 \cdot [\alpha_1 \cdot \lambda_0(\beta_1 \cdot t_2 + t)]$$
$$\text{for} \quad t \in (0, t_3 - t_2)$$
$$\lambda_{t_i}(t_i + t) = \prod_{k=1}^{i} \alpha_k \cdot \lambda_0 \left( \prod_{k=1}^{i} \beta_k \cdot t_k + t \right)$$
$$\text{for} \quad t \in (0, t_{i+1} - t_i)$$

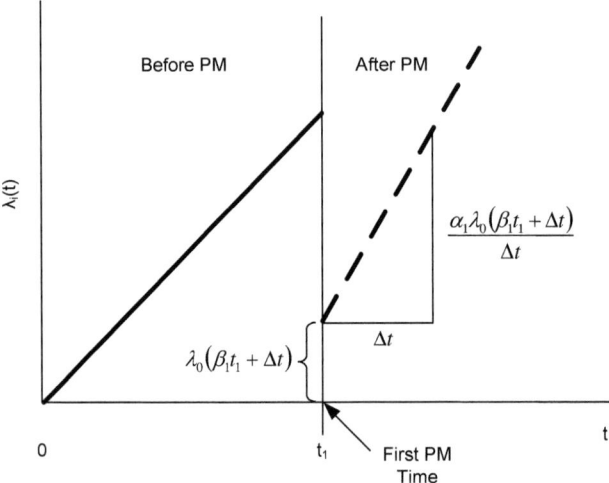

**Figure 4.3:** *Hybrid Model*

The benefit of this model is its ability to incorporate the advantages of the two preceding models. In the case of the failure rate preventive maintenance model, the rate of increase of the failure rate can be higher after each additional preventive maintenance task. Furthermore, the age reduction preventive maintenance model allows the determination of the instantaneous failure rate function value immediately after a preventive maintenance task ($\lambda(t) = \lambda(b_1 \cdot t_1)$ with $t_1 < t < t_2$).

One of the major drawbacks of this model is the absences of an explicit integration of maintenance quality levels and the loss of separation in maintainable and non-

## 4.2. Shape of the Failure Rate

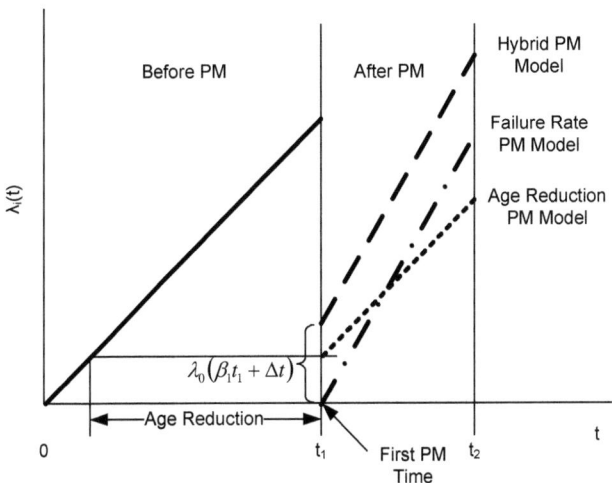

**Figure 4.4:** *Comparison of PM Models*

maintainable failure rates (see equation 4.1). If the constraints of the model ($1 = \alpha_0 \leq \alpha_1 \leq \ldots \leq \alpha_{N-1}, 0 = \beta_0 \geq \beta_1 \geq \ldots \geq \beta_{N-1} < 1$) were weakened, the quality levels could be integrated into the model and the failure rate could be splitten up into a part affected by preventive maintenance $\lambda_{Maint}(t)$ and an another, untouched part $\lambda_{Non-maint.}(t)$.

### 4.2 SHAPE OF THE FAILURE RATE

A multitude of different failure rate shape families have been proposed. Exponential distribution is one of the simplest distribution to model lifetimes and failure characteristics with its principle characteristics of being memoryless. The attractiveness mainly derives from its constant failure rate which allows a simple modelling. Although some failure characteristics do not follow this distribution, some sections of their failure rate are approximately constant.

A further developed family of distributions are monotone failure rates. They can be divided into increasing failure rates, modelling a deterioration, or decreasing failure rates representing an improvement of the system availability and reliability. Those monotone distributions are functions in time and are only capable to model some portions of the

lifetimes. They often fail to imitate the failure rate over the whole lifetime of an item.
Bathtub-shaped failure rate (see [Birolini, 2007]) is a more general failure rate shape, which has long been thought to represent a diversity of different lifetimes of components and systems. The shape starts with a decreasing failure rate, followed by a portion of constant failure rate and ends with an increasing failure rate. Those sections are related to infant mortality, middle lifetime and old age. Despite of its wide spread in the field of maintenance and availability modelling, its applicability limited. For example, Moubray [Moubray, 1991] showed, that only a marginal part of components in an aircraft follow a bathtub shape.

Failure rates are the mathematical representation of the adjoint failure mechanisms a given production system may be exposed to and describe the progressions of the failure mechanisms. Since the failure rate of a production system is the conglomerate of a multitude of different components and associated failure mechanisms, it would be self-evident to individually model every component and to condensate those failure mechanism related failure rates to one failure rate of the system. Such a modelling approach is provided by mixture models.

### 4.2.1 Mixture Models

$\lambda_{Failure}(t)$ is the outcome of a function incorporating all components and their possible failure mechanism. Whereas the effect of a single failure mechanisms on $\lambda_{Failure}(t)$ and the component's lifetime can roughly be estimated, the impact of all failure mechanisms and all components can hardly be quantified on system level. They mutually influence in the way that one failure mechanism promotes or impedes other failure mechanisms and the system failure rate cannot be understood as a simple aggregation of the individual failure rates. It is a mixture of mechanical and pneumatic failure rates, electronic failure rates and software failure rates. Figure 4.5 proposes an overview about the most important failure mechanisms.

Progression of all of these failure mechanisms can be described with different probability distributions. The combination of those different probability distributions can be depicted with mixture models. Those models are commonly related to a missing data problem where the sampled data points under consideration have *membership* in one of the distributions used to model the data. Either this *membership* is missing or the probability mixture model is used to describe system characteristics which emerge from the interplay of different probability distributions. In the later situation, defining the *membership* is part of the stochastic modelling and means to elaborate the kind of interaction

## 4.2. Shape of the Failure Rate

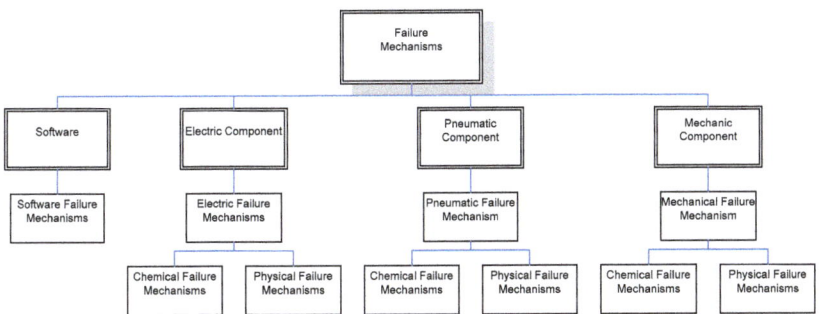

**Figure 4.5:** *Failure mechanisms*

of the probability distribution and to indicate the contribution of the single probability distributions by weightings $p_i$ [Pham, 2003].

Those different probability distributions arise because of two reasons; different failure mechanisms follow different probability distributions and not all of the identical components have exactly the same distribution (due to manufacturing tolerances). Some components have exponential distributed lifetimes whereas others failure mechanism is best described with a decreasing failure rate although they are exposed to the same failure mechanisms. Thus a mixture proceeding occurs automatically when different or even the same components are pooled. A mixture model represents a convex combination of independent, probability variables with eventually different distribution types [Finkelstein and Esaulova, 2001], [Wondmagegnehu, 2004]. An underlaying structure (failure mechanisms) is assumed so that each failure mode belongs to one or some number of different failure mechanisms. This failure mechanisms $\theta = 1..k$ may or may not be observable and can be represented in terms of their occurrence $f(t, \theta_i)$ and contribution $p_i$ to the system failure [Marin et al., 2005]. The $f(t, \theta_i)$'s are from potentially different parametric families with unknown parameter $\theta_i$.

According to Block et al. [Block and Savits, 2001], there are two ways of modelling the failure rates, additive and multiplicative (frailty model) models:

$$\text{Additive } \lambda(t, \theta_i) = \theta_i + \lambda(t)$$

and

$$\text{Multiplicative } \lambda(t, \theta_i) = \theta_i \cdot \lambda(t)$$

This $\lambda(t, \theta_i)$ can be used to derive:

$$f(t, \theta_i) = \text{Probability density function of } F(t, \theta_i)$$
$$= \lambda(t, \theta_i) \cdot e^{-\int_0^t \lambda(\gamma, \theta_i) d\gamma}$$
$$F(t, \theta_i) = \text{Failure distribution function with } P(T \leq t|\theta_i) = F(t, \theta_i)$$
$$\text{for the failure mechanism } \theta_i$$
$$p_i = \text{Weighting function/ Contribution of failure mechanism } \theta_i$$
$$\text{to mixing failure rate } \lambda_{Mixture}(t)$$

The failure rate $\lambda_{Mixture}(t)$ for the mixture is:

$$\lambda_{Mixture}(t) = \frac{f_{Mixture}(t)}{\int_t^\infty f_{Mixture}(\tau) d\tau}$$
$$f_{Mixture}(t) = \sum_{i=1}^{k} p_i \cdot f(t, \theta_i)$$

In this form of mixture, each of the failure mechanism is described by a component probability density function $f(t, \theta_i)$ and its mixture weight $p_i$. Weighting function $p_i$ expresses the probability that a failure comes from failure mechanism $\theta_i$ [Dey and Rao, 2005] (for further reading see [Barlow and Proschan, 1975] or [Navarro and Hernandez, 2004]).

Mixture models originate from descriptive statistics and constitute a novel way of analyzing and condensing data. Classically, they are used to describe complex systems and provide a framework for statistic modelling. There are two major fields of application for mixture models; either they are applied to recognize any structure in a given data set or they are directly used for stochastic modelling. Both applications are known in the maintenance subject.

In application, a major obstacle is the difficulty of estimation [Dey and Rao, 2005], which occurs at various levels:

1. **Parametric distribution family**. For any failure mechanism its appropriate failure probability function has to be selected. Even in the simplified case in which the $f(t, \theta_i)$'s are of the same distribution family, the right choice of the parametric family is a difficult task.

2. **Mutual impact of failure mechanisms and components**. Failure mechanisms and components do mutually affect and may even cause a shift between different

## 4.2. Shape of the Failure Rate

parametric families. A plain failure distribution function for a given failure mechanism can be indicated but this distribution function mutates at presence of other failure mechanisms. Even when the distribution family remains it is very likely that the parametric values of the distribution function will be altered.

3. **Parameter estimation**. Parameter estimation for $\theta_i$ is linked to the likelihood function

$$\mathbb{L}(\theta, p) = \prod_{l=1}^{n} \sum_{a=1}^{k} p_j f(t_l, \theta_a)$$

with $n$ observations and $i$ failure mechanisms. The explicit representation of the corresponding posterior expectations involves the expansion of the likelihood into $i^n$ terms which is too intricate to be used for more than a few observations. Using numerical optimization procedures like the Expectation - Maximization Algorithm to reduce computational complexity may help but can fail to converge to the major mode of the likelihood (weak convergence). Another difficulty is called "label switching" which causes nonsensical estimations. This effect is caused by a symmetry in the likelihood of the model parameters; the likelihood is the same for all permutations of the model parameters [Stephens, 2000].

Mixture models are an interesting, theoretic possibility to create the system failure rate $\lambda_{Mixture}(t)$ based on failure rates associated to the failure mechanisms and components, but lack of practical useability due to estimation shortcomings. Moreover, the applicability of the additive and multiplicative mixture models to represent system failure rate is questionable due to the estimation difficulties. Those models are only capable to depict linear dependencies between the different failure mechanisms. However, these interactions do not follow a linear approach, mostly (corrosion in combination with fraction is such an example).

### 4.2.2 Approximation of the Failure Rate $\lambda_{Failure}(t)$

It is known that the shape of the failure rate can hardly be directly derived from the underlaying failure mechanisms of the components, as shown in section 4.2.1. However, shapes of the failure rates are considered to be based on physical principles and on the heterogeneity of the "population" of components [Soyer et al., 2004].

Any shape of the failure rate can be created with those three distinctive curve portions:

- Descending failure rate (DFR)
- Constant failure rate (CFR)
- Increasing failure rate (IFR)

The fundamental question is which of those curve progressions, or combinations of basic progressions, approximate the failure rate of a production system best. On this account, systems with DFR, CFR and IFR are corrective and preventive maintained and their resulting availabilities are compared to verify or falsify the assumption that preventive maintenance is solitarily beneficial for systems with IFR (see [Gharbi et al., 2007], [Das et al., 2007]).

System availability $P_{0,P}(t)$ of a preventively maintained system can be calculated with a (Semi-) Markov chain (see Figure 4.7). Since $\lambda_{Failure}(t)[h^{-1}]$ is a function of $\lambda_{Prev.Maint}(t)[h^{-1}]$ with the characteristic that an increase in $\lambda_{Prev.Maint}(t)$ results in a decrease in $\lambda_{Failure}(t)$ and that $\mu_{Prev.Maint}(t)[h^{-1}] > \mu_{Repair}(t)[h^{-1}]$, system availability should have a maximum for $\lambda_{Prev.Maint}(t) > 0$ in the case of an increasing failure rate. $P_0(t)$, $P_1(t)$, and $P_2(t)$ are the state probabilities for being "Operating", in "Preventive Maintenance", and "Corrective Maintenance".

Above statement shall be proven with two models, one with corrective maintenance only (see Figure 4.6) and the other with both, preventive and corrective maintenance (see Figure 4.7). Let suppose the "Age Reduction Preventive Maintenance Model" (see subsection 4.1.2) to model impact of preventive maintenance on the failure rate.

$$\lambda_{Failure,i}(t) = \lambda_{Failure,i-1}(\beta_i \cdot t_i + t)$$
$$\text{for } t \in (0, t_{i+1} - t_i)$$

$\beta_i$ represents the impact of preventive maintenance on the failure rate and can be interpreted as effectiveness of a preventive maintenance activity to reduce the failure rate. Those two models are implemented in two different semi-Markov chains. $\lambda_{Prev.Maint}(t)$, $\mu_{Prev.Maint}(t)$ and $\mu_{Repair}(t)$ are:

$$\lambda_{Prev.Maint}(t) = b_0$$
$$\mu_{Prev.Maint}(t) = b_1$$
$$\mu_{Repair}(t) = b_2$$

## 4.2. Shape of the Failure Rate

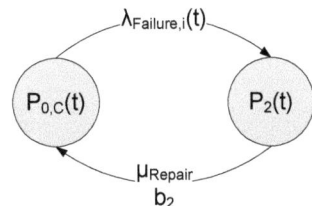

**Figure 4.6:** *Semi-Markov Chain of Corrective Maintenance*

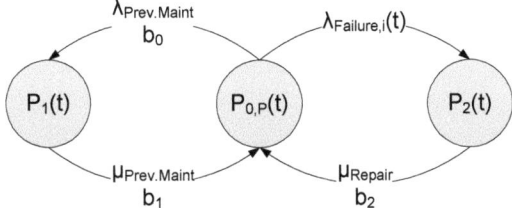

**Figure 4.7:** *Semi-Markov Chain of Preventive and Corrective Maintenance*

Those two Semi-Markov chains are implemented in the simulation environment $AnyLogic^{©}$ to investigate the effect of preventive maintenance in systems with different failure rate shapes on system availability.

### 4.2.3 Increasing Failure Rate

$\lambda_0(t)$ is the failure rate prior to the first failure and is given as:

$$\lambda_0(t) = m \cdot t + b$$
$$\text{for } t \in (0, t_1)$$

With $m$ representing the slope of the failure rate and $b$ the y-axis intersection. Then, failure rate $\lambda_i(t)$ is:

$$\lambda_{Failure,i}(t) = \lambda_{Failure,i-1}(\beta \cdot t_i + t)$$
$$\text{for } t \in (0, t_{i+1} - t_i)$$

The parameters of the semi-Markov chain were inspired by the values used in [Lin

**Chapter 4. Classic Maintenance Models and Failure Rate Shape**

et al., 2001] and are depicted in tables 4.1, 4.3 and 4.5. However, value estimation for the semi-Markov chain is less critical since only the difference of availability between the preventive an corrective maintained system is of interest.

For this experiment, two simulation runs are compared, one with $b_0 = 0.005$ and the other one with $b_0 = 0$ (no preventive maintenance). Figure 4.7 illustrates the system availability over time for corrective $P_{0,C}(t)$ and preventive maintenance $P_{0,P}(t)$.

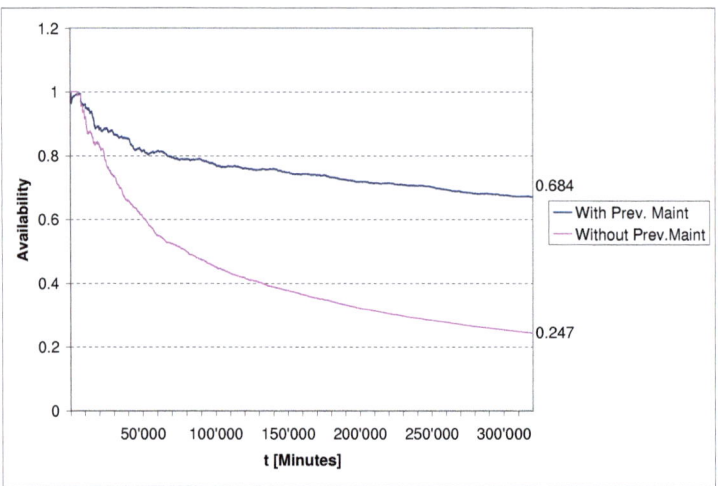

**Figure 4.8:** *Comparison between System Availability with and without Preventive Maintenance for Increasing Failure Rate*

Mission availability $A_{Mission_{S,C}}(t, k, T_i)[]$ for corrective and $A_{Mission_{S,P}}(t, k, T_i)[]$ for pre-

| | | |
|---|---|---|
| $b_0$ | 0.005 | $\left[\frac{1}{Minutes}\right]$ |
| $b_1$ | 0.08 | $\left[\frac{1}{Minutes}\right]$ |
| $b_2$ | 0.01 | $\left[\frac{1}{Minutes}\right]$ |
| $b$ | 0.0 | [] |
| $m$ | 0.0000003 | [] |
| $\beta$ | 0.9 | [] |

**Table 4.1:** *Parameters for the Markov Chain with Increasing Failure Rate*

## 4.2. Shape of the Failure Rate

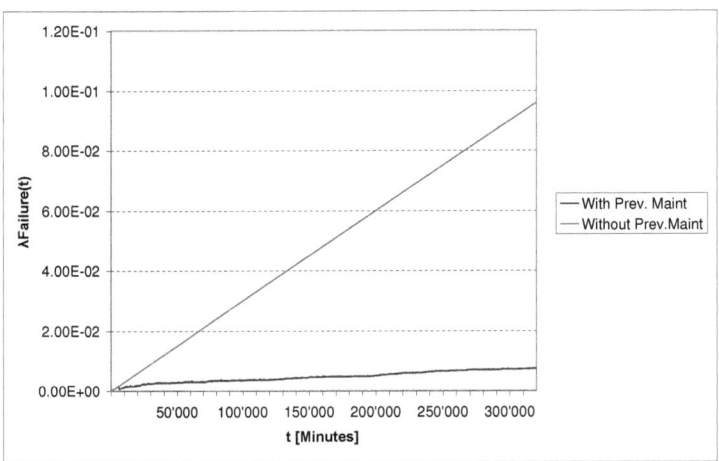

**Figure 4.9:** *Comparison between $\lambda_{Failure}(t)$ with and without Preventive Maintenance and Increasing Failure Rate*

ventive maintenance are defined as:

$$A_{Missions,C}(t,k,T_i) = \frac{1}{T}\int_{T_i} P_{0,C}(t)dt$$
$$A_{Missions,P}(t,k,T_i) = \frac{1}{T}\int_{T_i} P_{0,P}(t)dt$$

and mission time $T_i$ is set to 1'000 minutes. Results are shown in table 4.2
The following conclusions can be found:

- Dramatic gain of system availability with preventive maintenance in the long run (see Figure 4.8). This account is also given in the distribution of mission availabilities. Mean values of the mission availabilities improve from 0.247 to 0.684.

|  | With Prev. Maint. | Without Prev. Maint. |
| --- | --- | --- |
| $\mu(A_{Missions}(t,k,T_i))$ | 0.684 | 0.247 |
| $\sigma(A_{Missions}(t,k,T_i))$ | 0.030 | 0.044 |
| $CV[A_{Missions}(t,k,T_i)]$ | 0.253 | 0.849 |

**Table 4.2:** *Mean and Variance of Mission Availability with Increasing Failure Rate*

# Chapter 4. Classic Maintenance Models and Failure Rate Shape

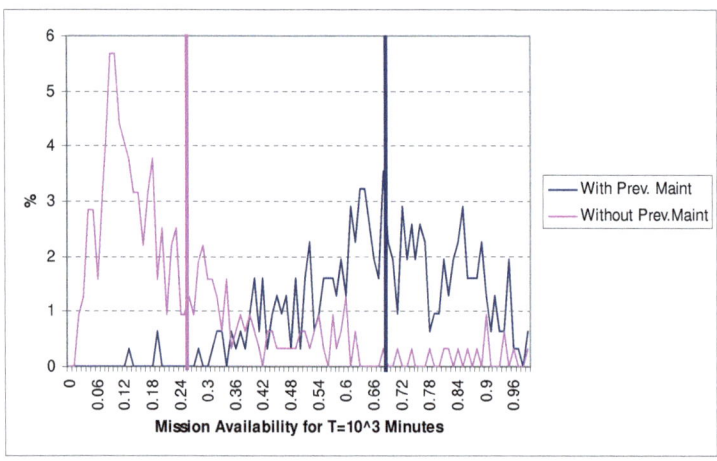

**Figure 4.10:** *Histogram of System Mission Availability with and without Preventive Maintenance and Increasing Failure Rate*

Availability improvement bases on a drastic lower failure rate increase than in the corrective maintenance case (see Figure 4.9).

- Not only the mean value of the mission availability distribution benefits from preventive maintenance but also the variances. This results in a drastic lower coefficient of variance. Coefficient of variance is used as a risk indicator. [Wellner, 2003]. Risk is associated with uncertainty, in this case, uncertainty in the mission availability. Uncertain mission availability increases the danger of non-deliverability and can only be compensated with higher safety factors (increasing planned lead-time) or stockkeeping. Since preventive maintenance impacts the coefficient of variance of the mission availability, it may help to fulfill the additional constraint to determine minimal planning horizon $T_{PH}$. Minimal planning horizon has to be elaborated in accordance with the terms that variation coefficient of demand $CV[P^{D(t,T_i)}]$ and mission availability distribution $CV[P^{A_{Mission_{S,P}}(t,k,T_i)}]$ must be lower or equal to 0.4 (compare with subsection 3.1.3). In this case, preventive maintenance reduces the coefficient of variance from 0.849 to 0.253 with a chosen mission time of 1'000 minutes. Thus, mission time could even be extended without violating the requirement of $CV[P^{A_{Mission_{S,P}}(t,k,T_i)}] \leq 0.4$.

## 4.2. Shape of the Failure Rate

- Preventive maintenance does not only affect system availability but also increases system output due to minimized safety factor and improves service level.

### 4.2.4 Constant Failure Rate

Constant failure rate exposes no temporal development and can be expressed as:

$$\lambda_{Failure,i}(t) = \lambda_{Failure,i-1}(t)$$
$$\text{for } t \in (0, t_{i+1} - t_i)$$

Since $\lambda_{Failure,i}(t)$ is not a function in time:

$$\lambda_{Failure,i}(t) = \lambda_{Failure,i-1}(t) = \lambda_0(t) = b$$

Corresponding parameters are defined in table 4.3

It can be stated that:

- Preventive maintenance has a slight impact on system availability in respect of long-term effect (see figure 4.11). System with preventive maintenance expose a marginal higher availability than the corrective maintained system. This anomaly is caused by model imperfection; a failure cannot occur when the system is in preventive maintenance. It is a problem of simultaneous events. Suppose the system is in preventive maintenance $P_1(t)$ and at the same time the transition from $P_{0,P}(t)$ to $P_2(t)$ is activated but cannot fire because the system is not in state $P_{O,P}(t)$. When the system is back in state $P_{0,P}(t)$ activation of the transition is already cleared in the Markov chain due to its memoryless behavior. In combination with the fact that a preventive maintenance action takes less time than a repair activity, system availability of preventive maintained systems may be higher

| | | |
|---|---|---|
| $b_0$ | 0.005 | $\left[\frac{1}{Minutes}\right]$ |
| $b_1$ | 0.08 | $\left[\frac{1}{Minutes}\right]$ |
| $b_2$ | 0.01 | $\left[\frac{1}{Minutes}\right]$ |
| $b$ | 0.003 | [] |
| $m$ | 0 | [] |
| $\beta$ | 0.9 | [] |

**Table 4.3:** *Parameters for the Markov Chain with Constant Failure Rate*

70    Chapter 4.  Classic Maintenance Models and Failure Rate Shape

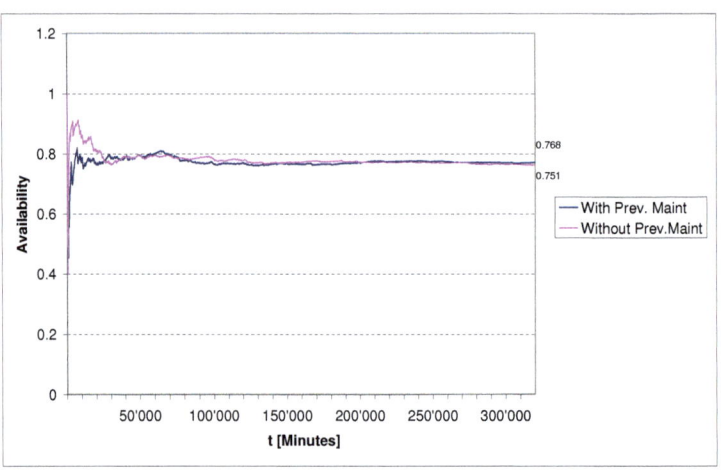

**Figure 4.11:** *Comparison between System Availability with and without Preventive Maintenance for Constant Failure Rate*

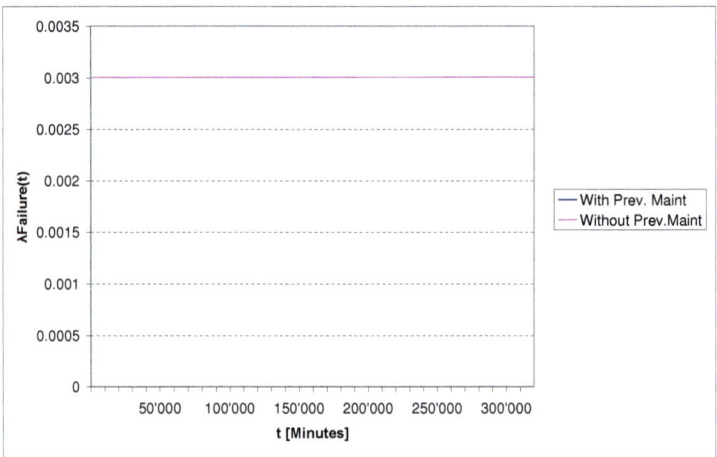

**Figure 4.12:** *Comparison between $\lambda_{Failure}(t)$ with and without Preventive Maintenance and Constant Failure Rate*

## 4.2. Shape of the Failure Rate

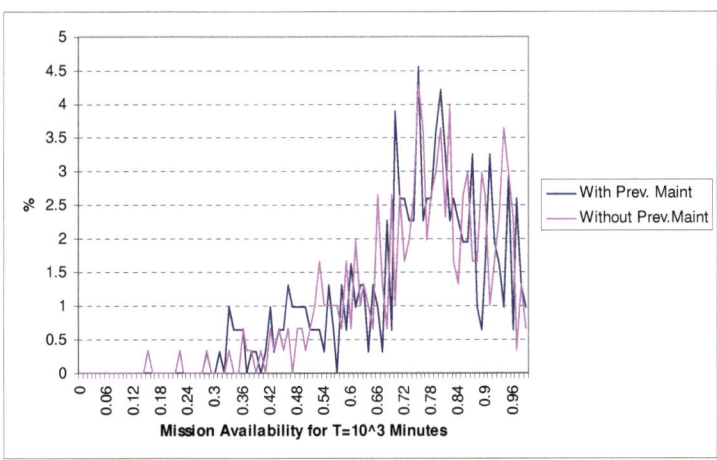

**Figure 4.13:** *Histogram of System Mission Availability with and without Preventive Maintenance and Constant Failure Rate*

|  | With Prev. Maint. | Without Prev. Maint. |
|---|---|---|
| $\mu(A_{Missions}(t,k,T_i))$ | 0.751 | 0.768 |
| $\sigma(A_{Missions}(t,k,T_i))$ | 0.028 | 0.026 |
| $CV[A_{Missions}(t,k,T_i)]$ | 0.223 | 0.210 |

**Table 4.4:** *Mean and Variance of Mission Availability with Constant Failure Rate*

than of corrective maintained systems even when the system expose a constant failure rate. However, this is only a modelling deficiency. This shortcoming could be overcome by introducing a fourth state representing the possibility of failure occurrence during preventive maintenance.

- In systems, whose failure characteristic is best described with a constant failure rate, preventive maintenance has a dual negative effect. First, system availability is reduced and second, needless preventive maintenance activities cause expenditures and seize capacities (compare with results in table 4.4).

- System availability of preventive and corrective maintained systems nearly coincide with increasing simulation time and approach a steady-state.

- There are only minor discrepancies concerning mission availabilities which arise from the different Markov chain setups. Thus, preventive maintenance impact on coefficient of variance of the mission availability distribution is neglectable (Figure 4.13).

### 4.2.5 Decreasing Failure Rate

Definition of this failure rate follows the explanations given in subsection 4.2.3. They differ only in the negative value for $m$ representing the slope of the failure rate. Associated parameters are shown in table 4.5

Regarding the definition of the decreasing failure rate in equation 4.2, the difficulty arises that:

$$\lambda_{Failure}(t) \geq 0$$

Taking the values given in the example, this condition is violated for $t > 10^6$.

- Preventive maintenance strongly affects system availability of systems with a decreasing failure rate as can be seen in Figure 4.11 and table 4.6. Whereas availability of the correctively maintained system constantly increases, availability of the other system stabilizes at a very low level. This stabilizing effect can be explained with the approximate constant failure rate of the preventively maintained system (see figure 4.16).

- It is the only case where preventive maintenance worsens the coefficient of variance of the mission availability distribution. Thus, preventive maintenance introduces uncertainty and destabilizes a system with decreasing failure rate.

- Preventive maintenance strategy has the most negative impact on systems with decreasing failure rate.

| | | |
|---|---|---|
| $b_0$ | 0.005 | $\left[\frac{1}{Minutes}\right]$ |
| $b_1$ | 0.08 | $\left[\frac{1}{Minutes}\right]$ |
| $b_2$ | 0.01 | $\left[\frac{1}{Minutes}\right]$ |
| $m$ | $-0.00000003$ | [] |
| $b$ | 0.03 | [] |
| $\beta$ | 0.9 | [] |

**Table 4.5:** *Parameters for the Markov Chain with Decreasing Failure Rate*

## 4.2. Shape of the Failure Rate

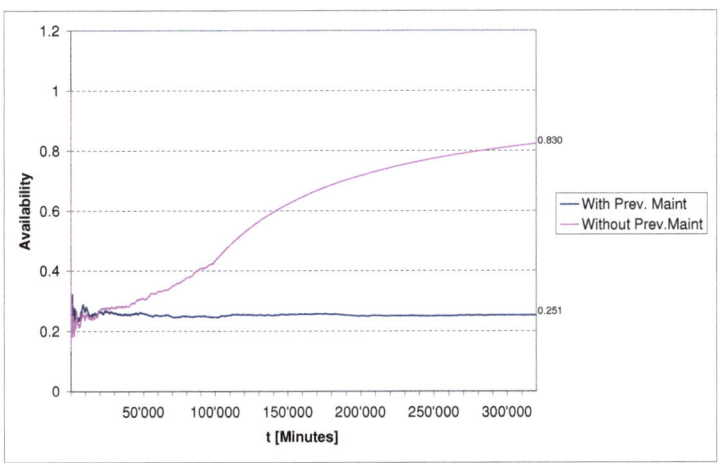

**Figure 4.14:** *Comparison between System Availability with and without Preventive Maintenance for Decreasing Failure Rate*

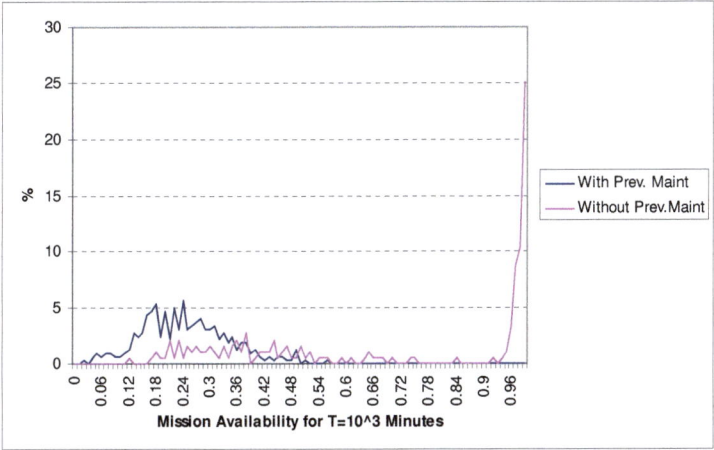

**Figure 4.15:** *Histogram of System Mission Availability with and without Preventive Maintenance and Decreasing Failure Rate*

# Chapter 4. Classic Maintenance Models and Failure Rate Shape

|  | With Prev. Maint. | Without Prev. Maint. |
|---|---|---|
| $\mu(A_{Missions}(t, k, T_i))$ | 0.251 | 0.830 |
| $\sigma(A_{Missions}(t, k, T_i))$ | 0.009 | 0.080 |
| $CV[A_{Missions}(t, k, T_i)]$ | 0.378 | 0.341 |

**Table 4.6:** *Mean and Variance of Mission Availability with Decreasing Failure Rate*

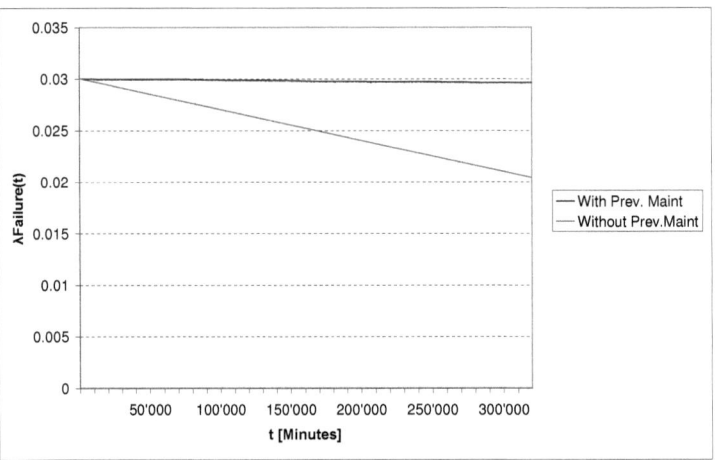

**Figure 4.16:** *Comparison between $\lambda_{Failure}(t)$ with and without Preventive Maintenance and Decreasing Failure Rate*

Preventive maintenance is the strategy of choice if the following two conditions are met:

- Maintained components exhibit an increasing failure rate (exhibit wearout) (see section 4.2.3)
- Overall preventive maintenance costs are lower than the cost of corrective maintenance

Simulation runs confirm that preventive maintenance strongly affects system availability only in the case of increasing failure rate. Provided that the model 4.7 is correct, an increase of system availability due to preventive maintenance is a strong indicator that the system failure rate must increase in time. If $\lambda_{Failure}(t)$ remains constant over

## 4.2. Shape of the Failure Rate

time, preventive maintenance would not substantially improve system availability and if $\lambda_{Failure}(t)$ even decreases, preventive maintenance would worsen system availability. Since many production systems are successfully maintained with preventive maintenance, it is reasonable to assume that $\lambda_{Failure}(t)$ is of the form:

$$\lambda_{Failure}(t) = m \cdot t + b \qquad (4.2)$$
$$m > 0$$
$$m \ : \quad \text{Slope}$$
$$b \ : \quad \text{Y-axis intercept}$$

Chapter 5

# Maintenance, Production and Logistic Model

Based on the hybrid preventive maintenance model by [Lin et al., 2001] an encompassing model is created (see Figure 5.1). This maintenance model is expanded to incorporate quality levels of maintenance activities and failed maintenance tasks. The whole model consists of four interconnected subsystems: *Maintenance Model*, *Production Model*, *Logistic Model* and *Main*. This separation provides easy extendability and offers the first time a simultaneous optimization of production, logistics, and production processes.

Core of the simulation is the *Maintenance Model* where the idea of improved hybrid preventive maintenance model is implemented (see section 5.1). This subsystem triggers the interruptions in the *Production Model* by message passing and calculates preventive maintenance and repair costs.

*Production Model* (section 5.3) represents the literal production process. As a first approximation, it is regarded as a black box without higher level of detail. System availability $A_{SS}(t)$ and coefficient of mission availability $CV\left[P^{A_{Mission_S}(k,p,T_i)}\right]$ are evaluated in the same model.

It provides the *Logistic Model* (see section 5.2) with the data stored in the object "ProductionOrder" (see 5.2.3.1). Service level $S_L(t,k)$, fill rate $F_R(t,k)$, coefficient of variation of the demand distribution $CV\left[P^{D(t,T_i)}\right]$, costs for delayed delivery $C_{Delayed}$, turnover $BV$ and material costs $C_{Mat}$ are calculated in the *Logistic Model*. This part of the simulation controls order dispatch and calculates logistic parameters. Planned finishing time $LT_j[h]$ is compared with actual production time $cLT_j[h]$ per order j to obtain the service level $S_L(t,k)$.

Discounted cash flow $DCF(t)$ and all costs $C_{Total}(t)$ are calculated in the *Main* section of the model (see section 5.4).

## 5.1 Maintenance Model

Zequeira [Zequeira and Bérenguer, 2005] and Lin [Lin et al., 2001] set the basis for the presented maintenance model (see Figure 5.2). It combines the hybrid preventive maintenance model with the idea of introducing quality levels [El-Ferik and Ben-Daya, 2006] for maintenance activities and the differentiation into maintainable $\lambda_{Maint}(t)\left[\frac{1}{h}\right]$ and non-maintainable $\lambda_{Non-Maint}(t)\left[\frac{1}{h}\right]$ failure rates [Lin et al., 2001]. The novelty of this model is the integration of all of these three ideas in one model. Principle feature of the model is its ability to model impairment of different maintenance activities on failure rate $\lambda_{Failure}(t)\left[\frac{1}{h}\right]$.

The dynamics of the model can be represented with the state chart in Figure 5.2. It incorporates three states:

- In Production
- In Preventive Maintenance
- In Failure/ Repair

"In Production" is a shared state with the *Production Model* (compare Figure 5.2 with Figure 5.22) since production process interacts with maintenance activities. State transitions connect the different states and fire after a given time $\frac{1}{\lambda(t)}$, respectively $\frac{1}{\mu(t)}$. Message passing to the *Production Model* assures that production is interrupted during maintenance or repair activities.

At the beginning of the simulation ($t = 0$), the maintenance module is in state "In Production". Depending on the values of $\lambda_{Failure}(t)$ and $\lambda_{Prev.Maint}(t)\left[\frac{1}{h}\right]$ either the transition to the failure state or to the preventive maintenance state fires. When the model is in one of those states, a stop signal "MessageDownTimeStart" is sent to the *Production Model* to interrupt production. Sojourn times in the appropriate state are recipro-

## 5.1. Maintenance Model

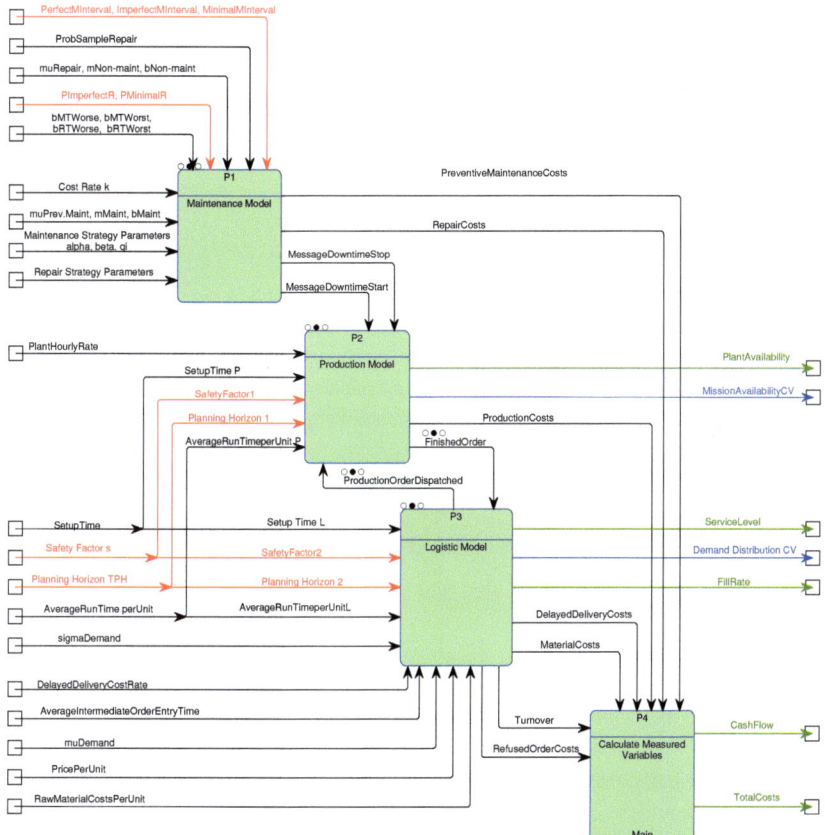

**Figure 5.1:** *Simulation Model with Target Values printed in green, Influence Parameters labelled in red and Constraints in blue*

cally proportional to the values of the transition rates from "In Preventive Maintenance" $\left(\mu_{Prev.Maint}(t)\left[\frac{1}{h}\right]\right)$ or "In Failure/ Repair" $\left(\mu_{Repair}(t)\left[\frac{1}{h}\right]\right)$ to the "In Production" state. Whenever the system enters "In Preventive Maintenance" or "In Failure/ Repair" state, the failure rate $\lambda_{Failure}(t)$ is recalculated. In the moment, when the system is leaving either state "In Preventive Maintenance" or "In Failure/ Repair", the message "Message-DownTimeStop" is passed to the *Production Model* to continue production.

# Chapter 5. Maintenance, Production and Logistic Model

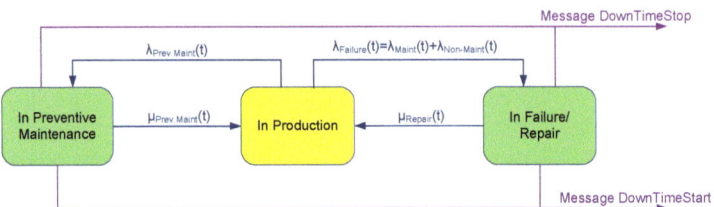

**Figure 5.2:** *Simplified Maintenance Model as a State Chart*

States "In Preventive Maintenance" and "In Failure/ Repair" are divided into quality levels. Impact of maintenance activities on the performance of a production systems (quality improvement, duration per piece, consumption of operating media, etc.) is not considered but could easily integrated by particularizing the state "In Production" according to maintenance states segmenting.

Basically, the *Maintenance Model* consists of four submodules (see Figure 5.3):

- *Preventive Maintenance Model* (see subsection 5.1.1)

- *Repair Model* (see subsection 5.1.2)

- *Failure Model* (see subsection) 5.1.3

- *Cost Calculation* (see subsection 5.1.4)

*Preventive Maintenance Model* and *Repair Model* provide the *Failure Model* with the variables $\alpha_{Prod}(k)$ and $\beta_{Prod}(k)$ to calculate the maintainable and non-maintainable part of the failure rate in the *Failure Model* (see subsections 5.1.1 and 5.1.2 for variable definitions and equation 5.11 for the definition of $\lambda_{Failure}(t)$). Whenever a repair or preventive maintenance activity starts, its impact on the failure rate $\lambda_{Failure}(t)$ (see Figure 5.2) is evaluated and the failure rate is updated in the *Failure Model*. When the system enters or leaves either *Preventive Maintenance Model* or *Repair Model*, message "MessageDowntimeStart" or "MessageDowntimeStop" is passed to the *Production Model* to interrupt production process. Sojourn times $T_{PM}, T_{IM}, T_{MM}, T_{WM}, T_{SM}, T_{PR}, T_{IR}, T_{MR}, T_{WR}$ and $T_{SR}$ in the associated quality levels, implemented in the *Preventive Maintenance Model* and *Repair Model*, are required in the *Cost Calculation* submodule to calculate the preventive maintenance costs $C_{Prev}$ and the repair costs $C_{Rep}$. The variables are defined in Figure 5.15.

## 5.1. Maintenance Model

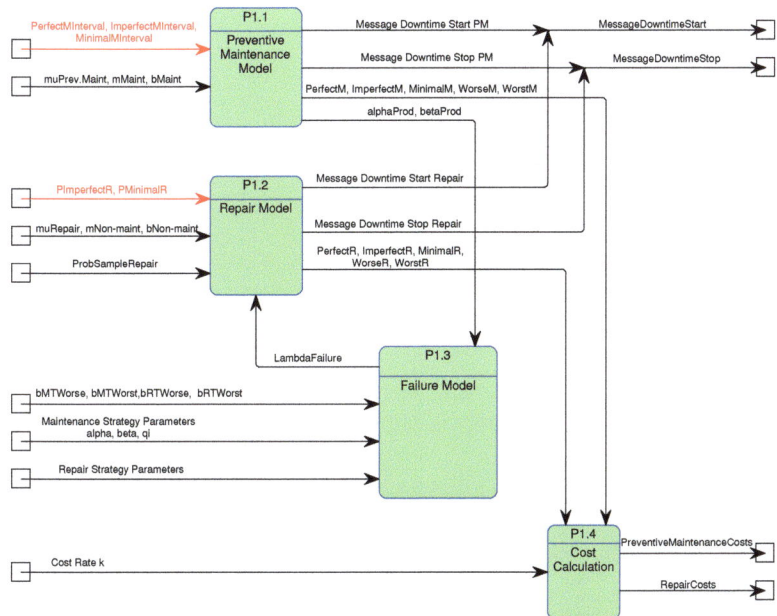

**Figure 5.3:** *Data Flow Diagram of the Maintenance Model*

### 5.1.1 Preventive Maintenance Model

Preventive maintenance state is split up into five sub-states representing the quality levels. According to the decision variables "PerfectMInterval", "ImperfectMinterval" and "MinimalMInterval"(see 5.1.1.3) and the failure rates $\lambda_{WorseM}(t) \left[\frac{1}{h}\right]$ and $\lambda_{WorstM}(t) \left[\frac{1}{h}\right]$ (see 5.1.2.3) maintenance model is set into the associated sub-state. Sojourn time in the sub-state is reciprocally dependent on the outgoing transition rate $\mu_{MT,i}(t) \left[\frac{1}{h}\right]$, whereas $i$ indicates the adjoint sub-state (see 5.1.2.4).

Entering one of the five preventive maintenance sub-states, factor $\alpha_i$ with co-domain $0 \leq \alpha_i \leq \infty$, representing its *long-term* impact on $\lambda_{Maint}(t)$.

$$\alpha_{Prod}(k) = \prod_{i=0}^{k} \alpha_i \qquad (5.1)$$

## Chapter 5. Maintenance, Production and Logistic Model

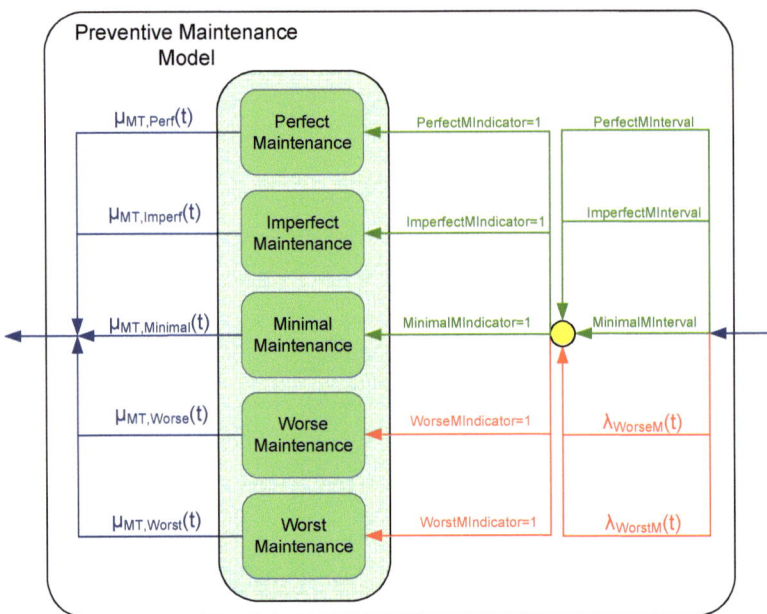

**Figure 5.4:** *Preventive Maintenance Model as a State Chart*

with

$\alpha_i$ : long-term impact of maintenance activity i on $\lambda_{Maint}(t)$

Table 5.1 gives reference about the $\alpha_i$ values. Their estimation is based on the following assumptions:

**Perfect Preventive Maintenance** This maintenance tasks set the failure rate back to the initial value ($\alpha_{Prod}(k) = 1$).

**Imperfect Preventive Maintenance** Gasmi et. al. [Gasmi et al., 2003] provide an estimate for the impact of imperfect preventive maintenance tasks of around $\alpha_i = 0.5$. Their approximation is based on a survey on hydro power systems and is used in this study due to missing real data.

**Minimal Preventive Maintenance** keeps the failure rate unaffected ($\alpha_i = 1$). [Worse Preventive Maintenance] Since the imperfect preventive maintenance quality level

## 5.1. Maintenance Model

**Preventive Maintenance Model**

| Influence Data | Symbol | Description | Codomain | Mean Value | Distribution | Data Origin |
|---|---|---|---|---|---|---|
| muPrev.Maint [1/h] | $\mu_{Prev.Maint}$ | Average rate between "Preventive Maintenance" and "In Production" state, is used to calculate the quality level specific transition rates | [0, ∞] | 0.333 | Exponential (according to Lit. Res.) | Company |
| $m_{Maint}$ [1/h] | $m_{Maint}$ | Slope of the maintainable failure rate | - | 0.0000099 | - | Estimation based on Company's Experience |
| $b_{Maint}$ [1/h] | $b_{Maint}$ | Y-axis interception of the maintainable failure rate | [0, ∞] | 0.0045 | - | Estimation based on Company's Experience |

| Decision Variables | Symbol | Description | Codomain | Start Value | Step Size |
|---|---|---|---|---|---|
| PerfectMInterval [h] | $I_{PM}$ | Time between two perfect preventive maintenance activities | [10'000, 60'000] | - | 10'000 |
| ImperfectMInterval [h] | $I_{IM}$ | Time between two imperfect preventive maintenance activities | [100, 10'000] | 100 | 100 |
| MinimalMInterval [h] | $I_{MM}$ | Time between two minimal preventive maintenance activities | [100, 1'000] | - | 100 |

| Output Data | Symbol | Description | Codomain |
|---|---|---|---|
| MessageDowntimeStop PM [] | - | Interrupts the production process during a preventive maintenance activity | [0, 1] |
| MessageDowntimeStart PM [] | - | Unblocks the production process after a preventive maintenance activity | [0, 1] |
| PerfectM [h] | $T_{PM}$ | Time spent for perfect maintenance | [0, ∞] |
| ImperfectM [h] | $T_{IM}$ | Time spent for imperfect maintenance | [0, ∞] |
| MinimalM [h] | $T_{MM}$ | Time spent for minimal maintenance | [0, ∞] |
| WorseM [h] | $T_{WM}$ | Time spent for worse maintenance | [0, ∞] |
| WorstM [h] | $T_{SM}$ | Time spent for worst maintenance | [0, ∞] |
| alphaProd [] | $\alpha_{Prod}(k)$ | Product of all α for k preventive maintenance activities | [0, ∞] |
| betaProd [] | $\beta_{Prod}(k)$ | Product of all β or k preventive maintenance activities | [0, ∞] |

**Figure 5.5:** *Variable Definitions for the Preventive Maintenance Model*

represents the counterpart to the worse preventive maintenance level, for this level the reciprocal value is chosen ($\alpha_i = 2$).

**Worst Preventive Maintenance** provokes an immediate failure ($\alpha_i = \infty$).

Lin et al. [Lin et al., 2001] propose to model the *instantaneous* impact of preventive

| Quality Level | Co-Domain | Value in Simulation |
|---|---|---|
| Perfect Preventive Maintenance | $\alpha_{Prod}(k) = 1$ | $\alpha_{Prod}(k) = 1$ |
| Imperfect Preventive Maintenance | $0 < \alpha_i < 1$ | 0.5 |
| Minimal Preventive Maintenance | $\alpha_i = 1$ | 1 |
| Worse Preventive Maintenance | $1 < \alpha_i < \infty$ | 2 |
| Worst Preventive Maintenance | $\alpha_i = \infty$ | $\infty$ |

**Table 5.1:** *Quality Levels of the PM Activities*

maintenance ($\beta_i$-factor) on the maintainable failure rate as:

$$\beta_i = \frac{i}{2 \cdot i + 1}$$

$i$ : Amount of performed preventive maintenance activities

The multiplication $\beta_{Prod}(k)$ of all $\beta_i$ values is passed to the *Failure Model* to reevaluate $\lambda_{Maint}(t)$ (see Figure 5.6).

$$\beta_{Prod}(i) = \prod_{i=0}^{k} \beta_i \qquad (5.2)$$

with

$\beta_i$ : short-term impact of maintenance activity i on $\lambda_{Maint}(t)$
$i$ : Amount of preventive maintenance activities

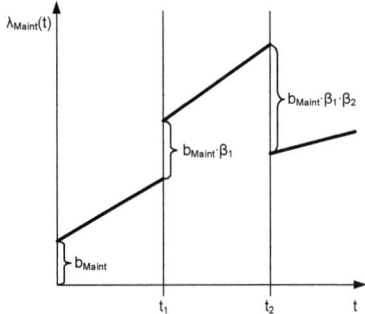

**Figure 5.6:** *Instantaneous Effect of Preventive Maintenance on Failure Rate*

#### 5.1.1.1 Value Estimation of $\mu_{Prev.Maint}$

Preventive maintenance time and system running time were recorded for a time span of one year (from 1.3.2006 till 28.2.2007 at Huba Control AG in Wuerenlos) to define a reliable estimate for $\mu_{Prev.Maint}$:

$$\mu_{Prev.Maint} = \frac{1}{\text{Mean Time for Preventive Maintenance}} = \frac{1}{3.00} = 0.333$$

The assumption about its probability distribution is based on literature research (see section 5.1.2.4 for literature).

## 5.1. Maintenance Model

### 5.1.1.2 Value Estimation of $\lambda_{Failure}(t)$

Following equation 4.2 the failure rate $\lambda_{Failure}(t)$ is of the form:

$$\lambda_{Failure}(t) = m \cdot t + b$$

with

$m$ : Slope
$b$ : Y-axis intercept

This failure rate can be estimated by using the average monthly failure rate $\lambda_{Failure}(t, T_i)$. Those values are taken from the same survey named in 5.1.1.1. Then, $\lambda_{Failure}(t)$ can be determined by a linear regression (see Figure 5.7).

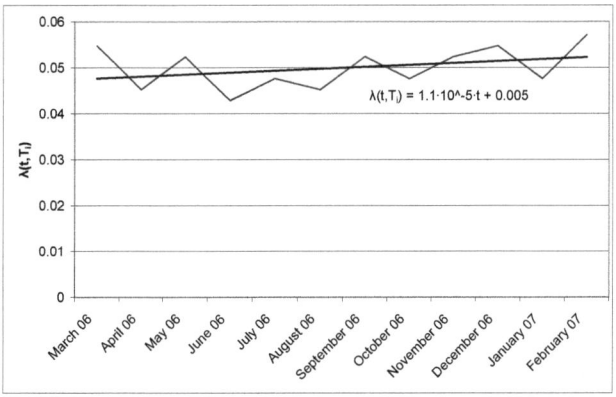

**Figure 5.7:** *Progression of the Failure Rate and Linear Approximation*

Linear approximation provides an estimation for the parameters $m$ and $b$:

$$m = 1.1 \cdot 10^{-5}$$
$$b = 0.005$$

Those values have to be subdivided into a maintainable and non-maintainable part. For this reason, an encompassing investigation about the possible and occurred failure modes of the production system has been conducted to decide whether a failure mode belongs to the maintainable part or not. This survey showed that most failure modes belong to the maintainable part (around 90%). Thus, values for the parameters of the maintainable and non-maintainable failure rates are:

$$\begin{aligned} m_{Maint} &= 1.1 \cdot 10^{-6} \cdot 0.9 = 0.99 \cdot 10^{-5} \\ b_{Maint} &= 0.005 \cdot 0.9 = 0.0045 \\ m_{Non-Maint} &= 1.1 \cdot 10^{-5} \cdot 0.1 = 0.11 \cdot 10^{-5} \\ b_{Non-Maint} &= 0.005 \cdot 0.1 = 0.0005 \end{aligned}$$

#### 5.1.1.3 Decision Variables in the Preventive Maintenance Model

In the adapted hybrid preventive maintenance model different maintenance strategies can be implemented by varying the maintenance intervals associated with the corresponding maintenance quality levels. Since *worse* and *worst* maintenance represent states of failed maintenance activities, which are out of control of the decision maker, only maintenance intervals for *perfect*, *imperfect* and *minimal* maintenance are at decision makers's disposal. *Worse* and *worst* maintenance are triggered by transition rates $\lambda_{WorseM}(t)$ and $\lambda_{WorstM}(t)$ that switch the reference variables "WorseMIndicator" and "WorstMIndicator" to 1 at $t = \frac{1}{\lambda_{WorseM}(t)}$ and $t = \frac{1}{\lambda_{WorstM}(t)}$ (see 5.1.2.2 for details). Decision variables trigger the timeout. At timeout, reference variables "PerfectMIndicator", "ImperfectMIndicator" or "MinimumMIndicator" are set to 1 and initiate the associated transition. This solution is chosen since several transitions to different preventive maintenance quality levels may fire simultaneously causing that diverse concurring events are in the event queue of the state chart. The simulation engine randomly selects one of the events in the event queue and all other events are cleared. Introduction of those reference variables provides accessability to these related events. To overcome randomness of event selection, a ranking procedure is introduced to list the events in a sequence according to their rating. The rating is:

1. Perfect Maintenance
2. Imperfect Maintenance
3. Minimal Maintenance
4. Worse Maintenance
5. Worst Maintenance

## 5.1. Maintenance Model

In the situation where the "Perfect Maintenance" and the "Imperfect Maintenance" transitions fire at the same time, the "Perfect Maintenance" transition will be favored. Outgoing transition rates of the preventive maintenance states model residence time in the associated state. Those residence times are exponential distributed and discussed in 5.1.2.4.

Whereas any failed maintenance activity may trigger a *worse* or *worst* maintenance, only depending on the severity of the failure, some distinctive maintenance tasks can be related to the other quality levels.

**Perfect Maintenance** A perfect maintenance activity requires a replacement of all wearout parts in a production system that $\lambda_{Maint}(t)$, $\lambda_{Non-maint}(t)$ are reset to zero. It is a complete overhauling or even a replacement of the production system.

**Imperfect Maintenance** Activities encompass cleaning, oiling, greasing, adjusting and so on. Some other actions are associated with the elimination of some weakspots of the production system. Thus, imperfect maintenance causes a partial modification of the system.

**Minimal Maintenance** The characteristic of a minimal maintenance activity is that the maintenance task does not impact the failure rate in the long run. This can be the replacement of a broken part by a used spare part with the same failure rate.

### 5.1.2 Repair Model

Although preventive maintenance is aimed at avoiding failures, some breakdowns cannot be avoided. After such a breakdown, a corrective maintenance task has to be performed to reset system into operation. These repair task affects $\lambda_{i,Maint}(t) \left[\frac{1}{h}\right]$ in the same way as described in the *Preventive Maintenance Model*.

After a failure has occurred, system is immediately set into repair state. The choice of which repair quality level is selected is taken accordingly to the quality level probability. Those probabilities are either directly determined by the decision variables $PImperfR$ and $PMinimalR$ (see Figure 5.9) or can be derived by the associated maintenance induced failure rates defined in 5.1.2.3. Maintenance failure probabilities $P_{WorseR}(t)$ and $P_{WorstR}(t)$ are [Birolini, 2007]:

# Chapter 5. Maintenance, Production and Logistic Model

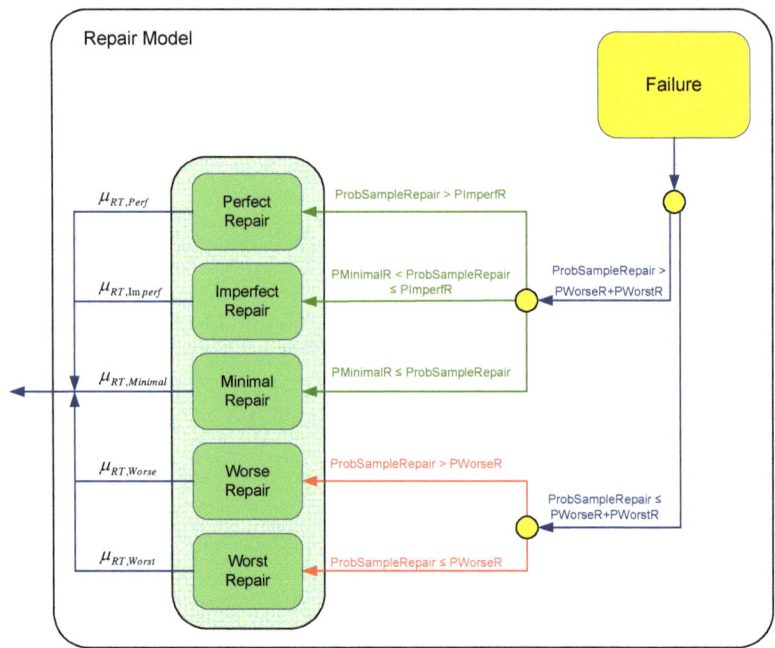

**Figure 5.8:** *Repair Model*

$$P_{WorseR}(t) = \lambda_{WorseR}(t) \cdot e^{-\int_0^t \lambda_{WorseR}(\tau)d\tau}$$
$$P_{WorstR}(t) = \lambda_{WorstR}(t) \cdot e^{-\int_0^t \lambda_{WorstR}(\tau)d\tau}$$

with

$\lambda_{WorseR}(t)$ : Failure rate for worse repair task
$\lambda_{WorstR}(t)$ : Failure rate for worst repair task

The decision variables $P_{ImperfR}$ and $P_{MinimalR}$ have to be chosen with respect to:

$$1 \geq P_{ImperfR} \geq P_{MinimalR} \geq P_{WorseR}(t) \geq P_{WorstR}(t) > 0$$

With those quality level related probabilities, a "decision network" can be built. Its systemic structure is shown in Figure 5.8. The randomized value of the uniform -distributed

## 5.1. Maintenance Model

**Repair Model**

| Influence Data | Symbol | Description | Codomain | Mean Value | Distribution | Data Origin |
|---|---|---|---|---|---|---|
| muRepair [1/h] | $\mu_{Repair}$ | Average rate between "Repair" and "In Production" state, is used to calculate the quality level specific transition rates | [0, ∞] | 0.345 | Exponential (according to Lit. Res.) | Company |
| $m_{Non-maint}$ [1/h] | - | Slope of the non-maintainable failure rate | - | 0.0000011 | - | Estimation based on Company's Experience |
| $b_{Non-maint}$ [1/h] | - | Y-axis interception of the non-maintainable failure rate | [0, ∞] | 0.0005 | - | Estimation based on Company's Experience |
| ProbSampelRepair [] | - | Uniform distributed probability variable used in the decision network to evaluate the appropriate quality level for repair | [0,1] | 0.5 | Uniform | - |

| Decision Variables | Symbol | Description | Codomain | Start Value | Step Size |
|---|---|---|---|---|---|
| PImperfectR [] | $P_{ImperfectR}$ | Probability indicator for imperfect repair | [$P_{MinimalR}$, 1] | 1 | 0.1 |
| PMinimalR [] | $P_{MinimalR}$ | Probability indicator for minimal repair | [$P_{WorseR}$, $P_{ImperfectR}$] | 0 | 0.1 |

| Output Data | Symbol | Description | Codomain |
|---|---|---|---|
| MessageDowntimeStop Repair [] | - | Interrupts the production process during a repair activity | [0, 1] |
| MessageDowntimeStart Repair [] | - | Unblocks the production process after a repair activity | [0, 1] |
| PerfectR [h] | $T_{PR}$ | Time spent for perfect repair | [0, ∞] |
| ImperfectR [h] | $T_{IR}$ | Time spent for imperfect repair | [0, ∞] |
| MinimalR [h] | $T_{MR}$ | Time spent for minimal repair | [0, ∞] |
| WorseR [h] | $T_{WR}$ | Time spent for worse repair | [0, ∞] |
| WorstR [h] | $T_{SR}$ | Time spent for worst repair | [0, ∞] |

**Figure 5.9:** *Variable Definition for the Repair Model*

probability variable $ProbSampleRepair[0,1]$ is compared with the transition conditions. When the comparison is evaluated to "true", the corresponding transition fires. Quality level probability is determined by the interval between two intersections on the x-axis (see figure 5.10).

### 5.1.2.1 Value Estimation of $\mu_{Repair}$

Corrective maintenance time and system running time have been recorded for a time span of one year (from 1.3.2006 till 28.2.2007 at Huba Control AG in Wuerenlos) to define a reliable estimate for $\mu_{Repair}$:

$$\mu_{Repair} = \frac{1}{MTTR} = \frac{1}{2.90} = 0.345$$

The assumption about its probability distribution is based on literature research (see section 5.1.2.4 for literature).

90                    Chapter 5. Maintenance, Production and Logistic Model

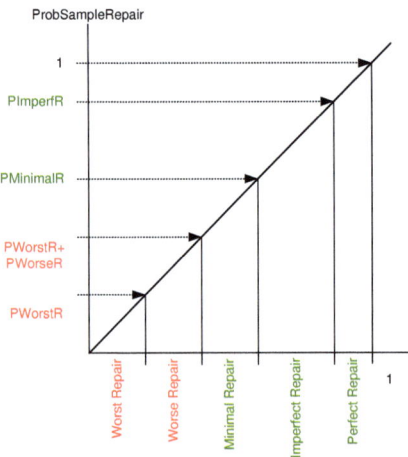

**Figure 5.10:** *Transformation of $ProbSampleRepair$*

### 5.1.2.2  Modeling Maintenance Induced Failures

In a study about preventive maintenance by United Airlines and American Airlines [Moubray, 1996], it was found that for a large class of rotating machines, the failure rate greatly increased just after the periodic overhauls - in other words, the overhaul reduced the reliability and availability of the machines. This effect is apparent if substantial parts of the system are replaced by new components and the shape of their failure rate follow a bath-tub-curve or they decrease in time. However, if this is not the case, sudden increasing in the failure rate after a preventive maintenance is addressed to failed maintenance. Human beings play an important role in maintenance and contribute to system failure by incorrect repair or preventive maintenance activity. A comprehensive overview about human error in maintenance can be found in [Dhillon and Liu, 2006].

Maintaining the wrong component, at the wrong point in time, or performing an inappropriate or inadequate maintenance activity can cause interruptions and are called maintenance induced failures. Some of the major causes for maintenance induced failures are:

- The need to disassemble some components to get access to other components, and removed components are often damaged [Crocker, 1999].

## 5.1. Maintenance Model

- Screw nuts and screws not adequately tightened and locked can loosen and if they are close to moving parts they can seriously damage the production system.
- Inappropriate lubricant or unauthorized spare parts are other sources of failures. Often, the latter are made of material of inferior quality or lack of production process quality. Useability of such fake spare parts may end in preterm component failure or damage.
- The latent danger of over-maintaining a production system characterized by performing preventive maintenance activities at more frequent intervals than necessary or executing tasks that are inefficient concerning adding value to the output [Anderson, 2002].

Maintenance-induced failures are related to human failures and their contribution to the entity of failures can be immense, raising up to 70% of all failures [Smith and Tate, 1998a]. Maintenance actions are included in the probabilistic safety assessment (PSA) but have never been in focus of developments in human reliability analysis. Some advancements have been made, particularly in assessing probabilities of human failures in maintenance [Reiman, 1994], [Swain and Guttmann, 1983] [Samanta et al., 1985], [Nelson, 1997] and [Pocock et al., 2001]. Although some approaches even try to simulate maintenance activities (MAPPS [Siegel et al., 1984]), developing and testing human reliability analysis models with real plant data has been done rarely.

Vaurio [Vaurio, 2001] developed some equations to estimate the failure probability of repeated maintenance activities. Those activities are modeled as a task cycle consisting of a sequence of n tasks. The failure probability of the $n+1^{th}$ task depends on the number of consecutive errors immediately preceding the action. After a success the failure probability in the following task is always the same and is independent of earlier results.

Sheu et. al. [Sheu et al., 2006] present a model for systems with maintenance in which imperfect, perfect and failed preventive maintenance are distinguished. Basically, the appropriate probabilities that preventive maintenance is perfect, imperfect, or failed, depend on the number of previously performed maintenance actions. This characteristic expresses the ability to learn and is achieved by integrating a learning curve approach. A version of an age reduction preventive maintenance model with different levels of repair (between a minimal repair to a perfect repair) is presented in Shirmohammadi et. al. [Shirmohammadi et al., 2007]. The idea was to use fixed cycle times between two consecutive preventive replacements. To prevent unnecessary replacements, a decision parameter is introduced to determine whether the system undergoes a replacement at the end of the cycle or the replacement is postponed until the next cycle

ends. However, Shirmohammadi et. al. limited the maintenance quality levels to minimal, imperfect and perfect maintenance.
Sánchez et. al. [Sánchez et al., 2006] presented a method to estimate preventive imperfect maintenance parameters. The effectiveness of a preventive maintenance activity and the linear aging rate are estimated with help of a maximum likelihood estimation (MLE). MLE is based upon a reliable database about maintenance activities, failure-reports and recorded production system availability.
Performing more frequent preventive maintenance tasks as necessary does not only waste resources but also increases possibility of maintenance-induced failures. Anderson [Anderson, 2002] states: "It cannot be assumed, however, that the benefit of carrying out the preventive maintenance activity increases system availability in all cases. In the case of intrusive preventive maintenance where assets are opened, adjusted, or otherwise handled, there is a chance that assets will be returned to service in a worse condition than when it was received. This means that as maintenance frequency increases the probability of a maintenance induced failure increases and the overall probability of success of the preventive maintenance activity is reduced (as demonstrated in Figure 5.11)". As maintenance frequency increases the likelihood of a maintenance-induced failure and the probability of failure prevention is reduced.

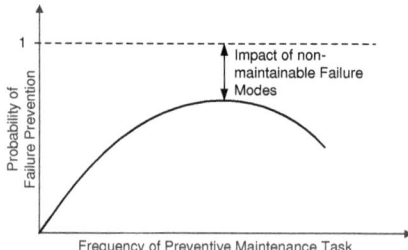

**Figure 5.11**: *Probability of Failure Prevention in Relation to Frequency of Preventive Maintenance Tasks according to [Anderson, 2002]*

Tracking maintenance-induced failures is very difficult and often a stringent connection between error and failed maintenance is absent due to the latency between failed maintenance and failure occurrence. Pekka [Pekka, 2000] performed a survey on maintenance-induced failures in nuclear a power plant that mainly expose two weaknesses:

1. **Limited to Nuclear Power Plants**. It is doubtful if the given results are applicable

## 5.1. Maintenance Model

to other nuclear power plants. However, it seems to be fairly impossible to transfer those results to production systems.

2. **Difficulty to derive State Transition Rates.** Absence of repair, maintenance and operating times make a state transition rate calculation impossible.

#### 5.1.2.3 Assumptions about Failed Maintenance and Repair Tasks

Literatur research highlights three tendencies concerning modelling maintenance and repair induced failures:

1. **Decreasing Failure Rate.** Decreasing rate of maintenance induced failures represents the training and experience effect of the maintenance crews. Since failure rates are exclusively defined for positive values, the first deviation of the failure rate function must be strictly negative (see [Vaurio, 2001], [Sheu et al., 2006] or [Shirmohammadi et al., 2007]).

2. **Increasing First Deviation of the Failure Rate.** A significant, positive correlation between maintenance performance and experience exist [Dhillon, 2002]. Thus, training and experience of maintenance follow a learning curve. With increasing training intensity effectiveness of training decreases (see the theory of the learning curve [Ritter and Schooler, 2002]). Since learning curve and failure rate of maintenance and repair induced failures expose opposed characteristics, first deviation of the failure is strictly negative (see [Ezey, 2000], [Riah-Belkaoui and Holzer, 1986] and [Sheu et al., 2006]).

3. **Codomain.** Codomain of the maintenance induced failure rate is between 0 and ∞ since this failure rate indicates a maintenance induced deterioration and not an improvement.

The time constant learning curve model provides all of those prerequisites (see [Wang and Lee, 2001] and compare with [Towill, 1990]) and is found to be adequate in most cases ( [Hackett, 1983]). The probability of a maintenance induced error $P_{MaintError}(t)$ is given as:

$$P_{MaintError}(t) = b - m\left(1 - e^{-\frac{t}{\tau}}\right) \quad (5.3)$$

$b$ : Initial failure probability
$\frac{m}{b}$ : Dynamic gain
$\tau[h]$ : Time constant in hours

Dillon [Dhillon, 2002] proclaimed a mean average human reliability of 0.987, meaning that one could expect probability error by maintenance personnel $P_{MaintError}(t)$ of 0.013. Company's experience shows that such a worst preventive maintenance is very seldom and is estimated to contribute with 5 to maximum 10% to failed maintenance (there is no documented worst preventive maintenance activity in the maintenance log-files of the company during the last 10 years).
Towill [Towill, 1990] proposed a time constant $\tau$ of 315 [h] and a relation of $\frac{b}{m} = 5.50$ for a switch assembling production system that is comparable to the situation at Huba Control AG.

$$P_{MaintError}(t) = P_{WorseM}(t) + P_{WorstM}(t)$$

With $P_{WorstM}(t) \approx 0.1 \cdot P_{WorseM}$:

$$P_{MaintError}(t) = 1.1 \cdot P_{WorseM}(t)$$
$$= 1.1 \cdot \left( b_{MTWorse} - \frac{1}{5.5} \cdot b_{MTWorse} \left(1 - e^{-\frac{t}{315}}\right) \right)$$

The assumption that $\frac{\int_0^{64'000} P_{MaintError}(t)dt}{64'000} = 0.013$ leads to the following equation:

$$0.013 = \frac{1.1}{64'000} \cdot \int_0^{64'000} \left( b_{MTWorse} - \frac{1}{5.5} \cdot b_{MTWorse} \left(1 - e^{-\frac{t}{315}}\right) \right) dt$$
$$b_{MTWorse} = 0.01440 \tag{5.4}$$

And $b_{MTWorst}$ can be computed with:

$$0.013 \cdot 0.1 = \frac{1}{64'000} \cdot \int_0^{64'000} \left( b_{MTWorst} - \frac{1}{5.5} \cdot b_{MTWorst} \left(1 - e^{-\frac{t}{315}}\right) \right) dt$$
$$b_{MTWorst} = 0.00159 \tag{5.5}$$

Experience shows that the probability failure occurrence increases with soaring stress level of the maintenance personnel and sinks with training the maintenance crews get (see [Kirwan, 1994], [Reason, 1990]). Repair activities are always unplanned interruptions in the production process and claim for immediate reaction to set production system back into operation. This instantaneous request of acting and repairing the production system under time constraints causes stress. In comparison with preventive maintenance tasks, experience in performing repair actions is lower than for preventive maintenance activities. That leads to the assumption that maintenance-induced failures are more likely for unplanned maintenance activities (repair, corrective maintenance) than for planned maintenance (preventive maintenance). Following the idea of

## 5.1. Maintenance Model

the AIPA model (compare with [Messen and Mohr, 1982]) which proposes an addition of 10% on the probability for the same activity under stress, it is assumed that:

$$b_{RTWorst} = 1.1 \cdot b_{MTWorst} = 0.00175 \quad (5.6)$$
$$b_{RTWorse} = 1.1 \cdot b_{MTWorse} = 0.01584 \quad (5.7)$$

Associated failure rate $\lambda_{MaintError}(t)$ for $P_{MaintError}(t)$ is (following [Mock, ]):

$$\begin{aligned}\lambda_{MaintError}(t) &= \frac{1}{1 - P_{MaintError}(t)} \cdot \frac{dP_{MaintError}(t)}{dt} \\ &= -\frac{0.000577 \cdot b \cdot e^{-\frac{t}{315}}}{1 - b + \frac{1}{5.5} \cdot b \cdot \left(1 - e^{-\frac{t}{315}}\right)}\end{aligned} \quad (5.8)$$

Equations 5.8 and 5.3 with the results in equations 5.4, 5.5, 5.6 and 5.7 provide the basis for the computation of all other failure rates and associated failure probabilities:

$$\lambda_{WorseM}(t) = -\frac{8.31 \cdot 10^{-6} \cdot e^{-\frac{t}{315}}}{0.988 - 0.0026 \cdot e^{-\frac{t}{315}}}$$
$$\lambda_{WorstM}(t) = -\frac{9.16 \cdot 10^{-7} \cdot e^{-\frac{t}{315}}}{0.987 - 0.0003 \cdot e^{-\frac{t}{315}}}$$
$$P_{WorseR}(t) = 0.01584 - 0.00288 \cdot \left(1 - e^{-\frac{t}{315}}\right)$$
$$P_{WorstR}(t) = 0.00175 - 0.00032 \cdot \left(1 - e^{-\frac{t}{315}}\right)$$

#### 5.1.2.4 Sojourn Times

In the maintenance model (see Figure 5.2), sojourn times in the states "Failure/ Corrective Maintenance" ($t_{RT}[h]$) and "Preventive Maintenance" ($t_{PM}[h]$) are reciprocal proportional to the outgoing transient rates $\mu_{Repair}(t) \left[\frac{1}{h}\right]$ and $\mu_{Prev.Maint}(t) \left[\frac{1}{h}\right]$. Literature proposes to model repair and maintenance times with an exponential probability distribution [Osaki, 2002], [Lewis, 1987], an Erlang probability distribution [Carmichael, 1987] or a lognormal distribution [Mi, 1991], [Birolini, 2007], [Ergam, 1982], [Almog, 1979], [Mullen, 2006].

Although lognormal distribution approximates repair and maintenance times best [Birolini, 2007], it is rarely applied due to the fact that calculations using this distribution tend to become very time-consuming. In the case, when the availability analysis of a system is in focus, lognormal distribution is well approximated by an exponential distribution with the same mean. This holds if ( [Birolini, 2007]):

$$MTTR(t + \Delta t) = MTTR(t)$$
$$MTTR \ll MTBF$$

Let assume that the system under consideration corresponds to the situation described by [Birolini, 2007]. Then, repair times $t_{RT}$ and preventive maintenance times $t_{PM}$ are exponential distributed. Since the quality levels of the maintenance tasks cannot be monitored, except for worst maintenance and worst repair, transition rates $\mu_{Repair}(t)$ and $\mu_{Prev.Maint}(t)$ are estimated by using the expected (average) residence times in the maintenance or repair state.

$$\mu_{Repair} = \frac{1}{E[t_{RT}]} \tag{5.9}$$

$$\mu_{Prev.Maint} = \frac{1}{E[t_{MT}]} \tag{5.10}$$

where

$$E[t_{RT}] = \frac{\text{Total Repair Time}}{\text{Sum of Repair Tasks}}$$

$$E[t_{MT}] = \frac{\text{Total Preventive Maintenance Time}}{\text{Sum of Preventive Maintenance Tasks}}$$

Since repair tasks cannot be prepared and unprepared activities normally take longer than prepared actions:

$$E[t_{RT}] > E[t_{MT}]$$

It is assumed, that repair and preventive maintenance tasks with a higher quality level require more time than an action with a lower quality level. Activities on a high quality level, perfect and imperfect repair, decrease failure rate of the system. Since they improve system availability, work content must contain some additional tasks which are absent in activities performed on lower quality levels. With the increase in work content, time to accomplish activity rises. By contrasting the minimal quality level with worse and worst quality level, the same analogy can be applied. Worse and worst quality level activities must either leave out some necessary tasks compared to the same activity on

## 5.1. Maintenance Model

minimal quality level or the wrong activities are performed. Let assume that the latter option is negligible. Those difference in the duration of the activities can be modelled by introducing task related quality factors, $q_i$, whereas $i$ represents the quality levels ($i$ =Perfect, Imperfect, Minimal, Worse, or Worst). Those quality factors $q_i$ are multiplied with $\mu_{Repair}$ and $\mu_{Prev.Maint}$. Repair and preventive maintenance related quality factors are defined as:

|  |  | Value |
|---|---|---|
| Perfect Repair/Preventive Maintenance | $q_{Perf}$ | 0.0024 |
| Imperfect Repair/Preventive Maintenance | $q_{Imperf}$ | 1 |
| Minimal Repair/Preventive Maintenance | $q_{Min}$ | 1 |
| Worse Repair/Preventive Maintenance | $q_{Worse}$ | 2 |
| Worst Repair/Preventive Maintenance | $q_{Worst}$ | 3.3 |

The values of the quality factors are approximated on the basis of the following approximations:

**Perfect Maintenance** $q_{Perf}$ A perfect repair or preventive maintenance task is related to a complete replacement of all parts exposing maintainable failure modes (e.g. all parts subject to wear-out). Such an overall replacement can be regarded as an overhaul of the whole production system and requires at least three months ($t = 21 \cdot 5 \cdot 3 = 1260[h]$). Thus, the residual time in the perfect repair and preventive maintenance state should be around 1260 hours.

$$q_{Perf} = \frac{\mu_{Prev.Maint}}{1260} = \frac{3}{1260} = 0.0024$$

**Imperfect Maintenance** $q_{Imperf}$ All performed preventive maintenance activities are aimed at decreasing the failure rate of the system. Therefore, the associated quality factor $q_{Imperf}$ reflects the actual required maintenance time and is set to 1.

**Minimal Maintenance** $q_{Min}$ A typical minimal maintenance activity is a replacement of a broken part by a spare part of the same age. Since the replacement time for a new or a used part does not substantially differ, it can be assumed that both replacement tasks are of the same duration and $q_{Min} \approx 1$.

**Worse Maintenance** $q_{Worse}$ Worse maintenance activities are tasks performed with inadequate accuracy due to stress, time pressure, carelessness or disregard of duty resulting in a saving of time. Thus, $q_{Worse}$ must be larger than 1. The saving of time is estimated to be in the range of 50% in comparison to the same task executed carefully ($q_{Worse} \approx \frac{1}{0.5} \approx 2$).

**Worst Maintenance** $q_{Worst}$ Worst maintenance is designated by insufficient preparation of the maintenance tasks and neglecting any direction, not reading the instructions, not following the instructions, using inadequate tools or spare parts and trying to save as much time as possible. According to Warnecke [Warnecke, 1992], the share of preparative operations is between 60-80% of the overall maintenance time. Let assume that a worst maintenance activity can be approximated by neglecting any preparative tasks and that they have an estimated share of 70% on the overall maintenance time, then $q_{Worst} \approx \frac{1}{1-0.7} \approx 3.3$

With help of those quality factors the transition rates can be expressed as:

$$\mu_{RT,i}(t) = q_i \cdot \mu_{Repair}(t)$$
$$\mu_{PM,i}(t) = q_i \cdot \mu_{Prev.Maint}(t)$$

with

$$i : Perf, Imperf, Min, Worse, Worst$$

### 5.1.3 Failure Model

Basically, *Failure Model* is responsible for the reevaluation of the failure rate $\lambda_{Failure}(t) \left[\frac{1}{h}\right]$ after a corrective or preventive maintenance activity.

$$\lambda_{Failure}(t) = \lambda_{k,Maint}(t) + \lambda_{j,Non-Maint}(t) \qquad (5.11)$$

#### 5.1.3.1 Derivation of $\lambda_{Maint}(t)$ and $\lambda_{Non-maint}(t)$

Definition of $\lambda_{1,Maint}(t) \left[\frac{1}{h}\right]$ follows the hybrid preventive maintenance model, as defined in 4.2, and describes the failure rate between the first (at $t = t_1$) and the second (at $t = t_2$) preventive maintenance tasks.

$$\lambda_{1,Maint}(t) = \alpha_1 \cdot m_{Maint} \cdot t + \beta_1 \cdot b_{Maint}$$

where

$m_{Maint}$ : Slope of the maintainable failure rate
$b_{Maint}$ : Y-Axis intercept of the maintainable failure rate
$\alpha_1$ : Quality level of the first preventive maintenance task
$\beta_1$ : Immediate effect of the first preventive maintenance task

## 5.1. Maintenance Model

The general form of the *maintainable failure rate* $\lambda_{k,Maint}(t)$ after the $k^{th}$ preventive maintenance action is:

$$\lambda_{k,Maint}(t) = t \cdot m_{Maint} \cdot \prod_{i=0}^{k} \alpha_i + b_{Maint} \cdot \prod_{i=0}^{k} \beta_i \qquad (5.12)$$

$\alpha_i$ : Quality level of the $i^{th}$ preventive maintenance task
$\beta_i$ : Immediate effect of the $i^{th}$ preventive maintenance task on the maintainable failure rate
$k$ : $k^{th}$ preventive maintenance action

With $\alpha_{Prod}(k)$ (see equation 5.1) and $\beta_{Prod}(k)$ (see equation 5.2) equation 5.12 can be simplified:

$$\lambda_{k,Maint}(t) = t \cdot m_{Maint} \cdot \alpha_{Prod}(k) + b_{Maint} \cdot \beta_{Prod}(k) \qquad (5.13)$$

The non-maintainable part of the failure rate follows the same idea as presented for the maintainable failure rate. Equivalent equation for $\lambda_{Non-Maint}(t) \left[\frac{1}{h}\right]$ is:

$$\lambda_{Non-Maint}(t) = t \cdot m_{Non-Maint} + b_{Non-Maint} \qquad (5.14)$$

with

$m_{Non-Maint}$ : Slope of the non-maintainable failure rate
$b_{Non-Maint}$ : Y-Axis intercept of the non-maintainable failure rate

Formula 5.12 and 5.14 can be inserted into equation 5.11 to receive the combined failure rate $\lambda_{Failure}(t)$. Temporal progression of $\lambda_{Failure}(t)$ before and after the second preventive maintenance action is depicted in figure 5.12. Encircled numbers from 0 to 6 indicate corrective maintenance actions; actually, they mark the end of the corrective maintenance activity. Failure rates are increasing during a maintenance action since they are independent from the state of the system but are an ordinary function in time. When the maintenance activity is finished, the new failure rate is evaluated.

All variables of the failure model are summarized in Figure 5.13 and Figure 5.14 represents the whole maintenance model with its individual sub-models *Preventive Maintenance Model*, *Repair Model* and *Failure Model*.

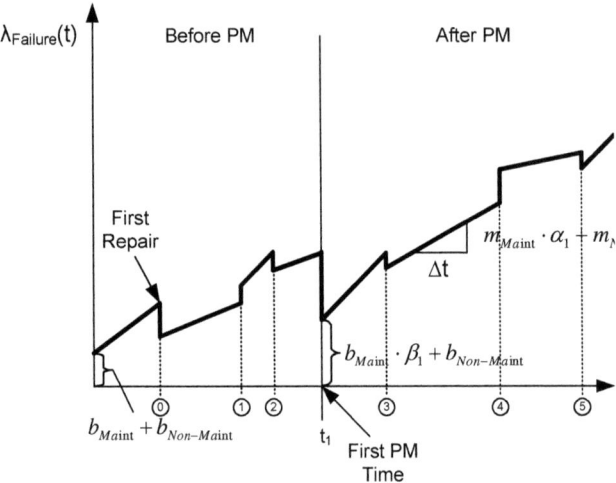

**Figure 5.12:** *Qualitative Failure Rate Progression before and after the first Preventive Maintenance Activity*

## 5.1. Maintenance Model

**Failure Model**

| Influence Data | Symbol | Description | Codomain | Mean Value | Distribution | Data Origin |
|---|---|---|---|---|---|---|
| τ [h] | - | Time constant | - | 315 | - | Towill1990 |
| $b_{MTWorse}$ [] | - | Y-axis interception of the maintenance induced failure rate for worse maintenance | - | 0.0144 | - | Calculation |
| $b_{MTWorst}$ [] | - | Y-axis interception of the maintenance induced failure rate for worst maintenance | - | 0.00159 | - | Calculation |
| $b_{RTWorse}$ [] | - | Y-axis interception of the maintenance induced failure rate for worse repair | - | 0.01584 | - | Calculation |
| $b_{RTWorst}$ [] | - | Y-axis interception of the maintenance induced failure rate for worst repair | - | 0.00175 | - | Calculation |
| $\alpha_{Imperf}$ [] | - | α-value exposes the impact of imperfect maintenance on the maintainable failure rate slope | - | 0.5 | - | Gasmi2003 |
| $\alpha_{Minimal}$ [] | - | α-value exposes the impact of minimal maintenance on the maintainable failure rate slope | - | 1 | - | ElFe2006 |
| $\alpha_{Worse}$ [] | - | α-value exposes the impact of worse maintenance on the maintainable failure rate slope | - | 2 | - | Gasmi2003 |
| $\alpha_{Worst}$ [] | - | α-value exposes the impact of worst maintenance on the maintainable failure rate slope | - | ∞ | - | ElFe2006 |
| $\beta_i$ [] | - | Immediate effect of preventive maintenance on the maintainable failure rate (i=Amount of prev.maint activities) | [0, 0.5] | | $\beta_i = i/(2i+1)$ | Lin2001 |
| alphaProd [] | $\alpha_{Prod}(k)$ | Product of all α for k preventive maintenance activities | [0, ∞] | - | - | - |
| betaProd [] | $\beta_{Prod}(k)$ | Product of all β or k preventive maintenance activities | [0, ∞] | - | - | - |
| $q_{Perf}$ | - | Quality factor for perfect maintenance | - | 0.0024 | - | Calculation |
| $q_{Imperf}$ | - | Quality factor for imperfect maintenance | - | 1 | - | Calculation |
| $q_{Minimal}$ | - | Quality factor for minimal maintenance | - | 1 | - | Calculation |
| $q_{Worse}$ | - | Quality factor for worse maintenance | - | 2 | - | Calculation |
| $q_{Worst}$ | - | Quality factor for worst maintenance | - | 3.3 | - | Calculation |

**Figure 5.13:** *Variable Definition of the Failure Model*

# Chapter 5. Maintenance, Production and Logistic Model

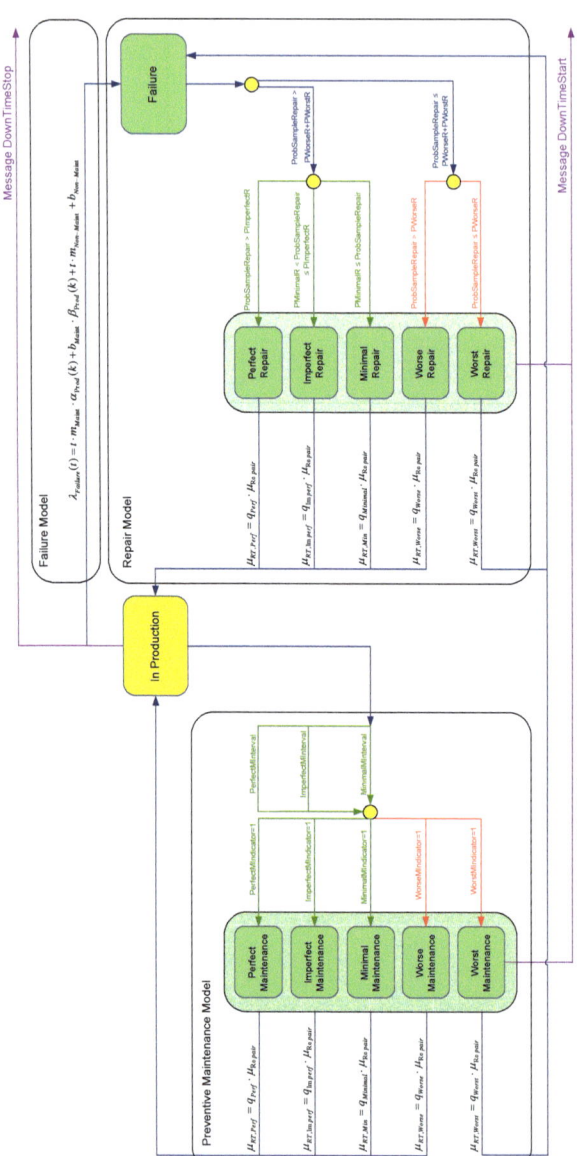

**Figure 5.14:** *Maintenance Model consists of the Adapted Hybrid Preventive Maintenance Model (see figure 5.4), the Repair Model 5.8 and the Failure Model 5.12*

## 5.2. Logistic Model

### 5.1.4 Cost Calculation

Costs associated with performed maintenance tasks are considered to be linear dependent on the duration of the maintenance task [Zequeira and Bérenguer, 2005], or being fix [Lin et al., 2001]. In this model, the sojourn time in the maintenance or repair state is multiplied with a fix cost factor $k$ to obtain the maintenance and repair costs:

$$C_{Prev} = k \cdot (T_{PM} + T_{IM} + T_{MM} + T_{WM} + T_{SM}) \quad (5.15)$$

$$C_{Rep} = k \cdot (T_{PR} + T_{IR} + T_{MR} + T_{WR} + T_{SR}) \quad (5.16)$$

The variables are defined in table 5.15.

**Cost Calculation**

| Influence Data | Symbol | Description | Codomain | Mean Value | Distribution | Data Origin |
|---|---|---|---|---|---|---|
| CostRate [sFr./h] | k | Cost rate per hour for the idle production system, used to calculate idle costs | $[0, \infty]$ | 220 | - | Company |
| PerfectR [h] | $T_{PR}$ | Time spent for perfect repair | $[0, \infty]$ | 0 | - | - |
| ImperfectR [h] | $T_{IR}$ | Time spent for imperfect repair | $[0, \infty]$ | 0 | - | - |
| MinimalR [h] | $T_{MR}$ | Time spent for minimal repair | $[0, \infty]$ | 0 | - | - |
| WorseR [h] | $T_{WR}$ | Time spent for worse repair | $[0, \infty]$ | 0 | - | - |
| WorstR [h] | $T_{SR}$ | Time spent for worst repair | $[0, \infty]$ | 0 | - | - |
| PerfectM [h] | $T_{PM}$ | Time spent for perfect maintenance | $[0, \infty]$ | 0 | - | - |
| ImperfectM [h] | $T_{IM}$ | Time spent for imperfect maintenance | $[0, \infty]$ | 0 | - | - |
| MinimalM [h] | $T_{MM}$ | Time spent for minimal maintenance | $[0, \infty]$ | 0 | - | - |
| WorseM [h] | $T_{WM}$ | Time spent for worse maintenance | $[0, \infty]$ | 0 | - | - |
| WorstM [h] | $T_{SM}$ | Time spent for worst maintenance | $[0, \infty]$ | 0 | - | - |
| **Output Data** | **Symbol** | **Description** | **Codomain** | | | |
| PreventiveMaintenanceCosts [sFr.] | $C_{Prev}$ | $C_{Prev} = (T_{PM}+T_{IM}+T_{MM}+T_{WM}+T_{SM}) \cdot k$ | $[0, \infty]$ | | | |
| RepairCosts [sFr.] | $C_{Rep}$ | $C_{Rep} = (T_{PR}+T_{IR}+T_{MR}+T_{WR}+T_{SR}) \cdot k$ | $[0, \infty]$ | | | |

**Figure 5.15:** *Variable Definition for Cost Calculation in the Maintenance Model*

## 5.2 LOGISTIC MODEL

A multitude of attempts combining maintenance strategies with logistics and manufacturing concepts have been done. Pinjala et al. [Pinjala et al., 2006] investigated the relationship between business and maintenance strategies with an empirical survey. They highlighted the maintenance link with business strategy elements by studying the impact of being *cost competitor*, *quality competitor* or *flexibility competitor* on the related maintenance strategy. In 2001, Perkins et al. [Perkins and Srikant, 2001] introduced a model of a failure-prone production system with uncertain demand to estimate the optimal hedging point policy; policy for with the production system is operated at maximum capacity until the buffer reaches a certain level, the hedging point. Then, the production system is turned off. Demand arrives according to a Poisson Process and burst size is exponential distributed. Incoming orders are immediately processed on a buffered production system with exponential distributed up- and downtimes. Some other papers are related to the problem of optimal number of Kanban cards in failure-prone production systems (see [Abdulnour et al., 1995], [Albino et al., 1992] and [Savsar, 1997]. Kanban system is a strategy for material management in logistics with continuous demand aimed at minimizing safety stock and material in process. Two consecutive departments build a Kanban circle in which a fix amount of Kanban cards (orders) circulate and production starts when a Kanban card arrives at the preceding department (for further information please see [Schoensleben, 2002]). However, since the production system in focus is implemented in a JiT-logistics but not in a Kanban system, these sophisticated models are out of importance.

Kenné et al. [Kenné et al., 2006] studied the effect of different maintenance and manufacturing strategies under the constraint of lost sales due to system unavailability. In comparison to the following model, they incorporated inventory. The proposed model is geared to the demand modelling presented by Pinjala et al. [Pinjala et al., 2006], but demand distribution is chosen according to real data of the partner company and integrates the idea of lost sales [Kenné et al., 2006].

*Service Level Determination, Order Dispatching* and *Production Calculation* are executed in the *Logistic Model*. Mainly, this part of the simulation is responsible for providing the *Production Model* with production orders, which are routed as "Orders" from the *Logistic Model* to the *Production Model*.

## 5.2.1 Service Level Determination

Planned lead time of a production order is compared with actual lead time in the service level section. When a finished production order enters the *Logistic Model*, planned finishing time $cLT_j$ and order finishing time $LT_j$ are compared and the boolean equation

$$LT_j > cLT_j$$

is evaluated. Depending on the result of this equation, the production order is in time or out of time.

$$S_L(t,k) = 1 - \frac{O_{Delayed}(k)}{P_O(k)} \qquad (5.17)$$

## 5.2. Logistic Model

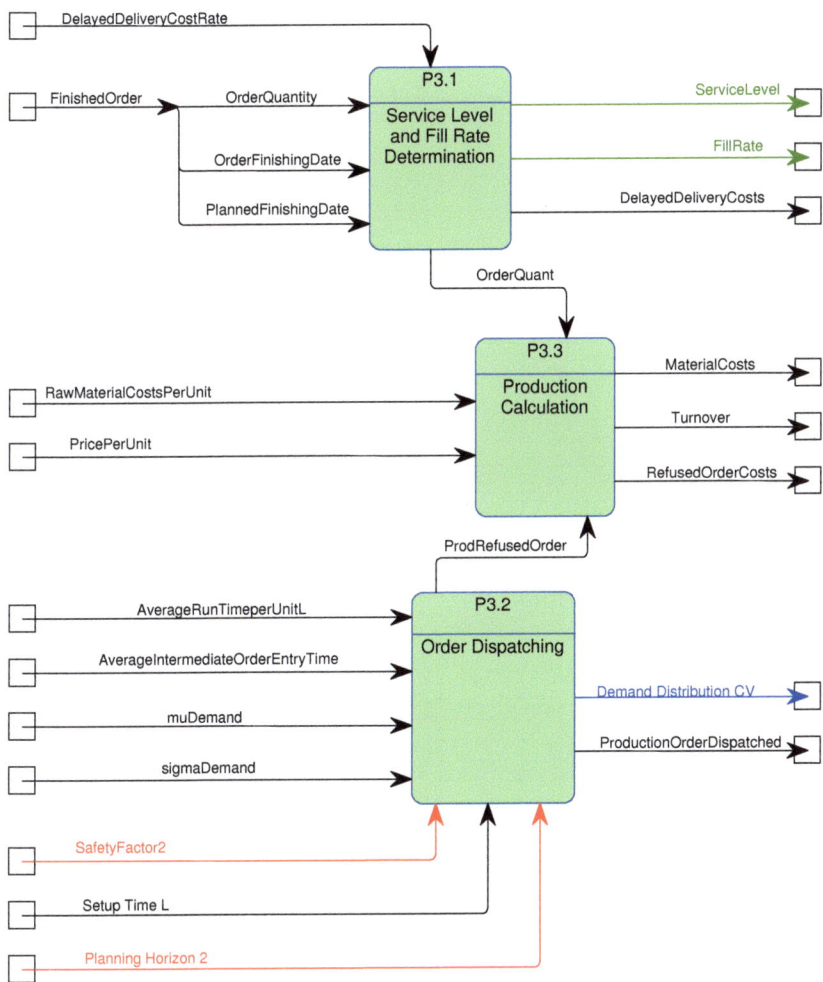

**Figure 5.16:** *Logistic Model*

# Chapter 5. Maintenance, Production and Logistic Model

## Logistic Model

### Service Level and Fill Rate Determination

| Influence Data | Symbol | Description | Codomain | Mean Value | Distribution | Data Origin |
|---|---|---|---|---|---|---|
| DelayedDeliveryCostRate [sFr./Order] | $CR_{Delayed}$ | Cost rate for delayed delivery | $[0, \infty]$ | 1'000 | - | Company |
| OrderQuant [pc.] | $Q_{Order}$ | Random order quantity following the distribution $Q(\mu_{DemandQuantity}, \sigma_{DemandQuantity})$ | $[0, \infty]$ | 2'134.9 | Weibull | Company |
| OrderFinishingDate [h] | $LT_j$ | Lead time for order j | $[0, \infty]$ | - | - | Company |
| PlannedFinishingDate [h] | $cLT_j$ | $= (Q_{Order} \cdot CT + T_{Setup}) \cdot s_T$ | $[0, \infty]$ | - | - | Company |

| Auxiliary Variables | Symbol | Description | Codomain | Mean Value | Distribution |
|---|---|---|---|---|---|
| DelayedOrders | $O_{Delayed}(k)$ | Amount of delayed orders of k orders | | | |
| DelayedOrderQuantities | $Q_{Delayed}(k)$ | Amount of delayed quantities | | | |
| ProductionVolume | $P_V(k)$ | Amount of production until $k^{th}$ order | | | |
| ProductionOrders | $P_O(k)$ | Amount of order until $k^{th}$ order | | | |

| Output Data | Symbol | Description | Codomain | Start Value | Distribution |
|---|---|---|---|---|---|
| ServiceLevel [] | $S_L(t,k)$ | $S_L(t,k) = 1 - O_{Delayed}(k)/P_O(k)$ | $[0, 1]$ | - | - |
| FillRate [] | $F_R(t,k)$ | $F_R(t,k) = 1 - Q_{Delayed}(k)/P_V(k)$ | $[0, 1]$ | - | - |
| DelayedDeliveryCosts [sFr.] | $C_{Delayed}$ | $C_{Delayed} = O_{Delayed}(k) \cdot CR_{Delayed}$ | $[0, \infty]$ | - | - |
| OrderQuant [pc.] | $Q_{Order}$ | Random order quantity following the distribution $Q(\mu_{DemandQuantity}, \sigma_{DemandQuantity})$ | $[0, \infty]$ | 2'134.9 | Weibull |

### Production Calculation

| Influence Data | Symbol | Description | Codomain | Mean Value | Distribution | Data Origin |
|---|---|---|---|---|---|---|
| PriceperUnit [sFr.] | $P_{PU}$ | Average retail price per unit | $[0, \infty]$ | ■ | - | Company |
| RawMaterialCostsperUnit [sFr.] | $C_{RAW}$ | Average raw material costs per unit | $[0, \infty]$ | ■ | - | Company |
| OrderQuant [pc.] | $Q_{Order}$ | Random order quantity following the distribution $Q(\mu_{DemandQuantity}, \sigma_{DemandQuantity})$ | $[0, \infty]$ | 2'134.9 | Weibull | Company |
| ProdRefusedOrder [pc.] | $P_{RO}$ | Sum of the volume of all refused orders | $[0, \infty]$ | - | - | - |

| Output Data | Symbol | Description | Codomain | Start Value | Distribution |
|---|---|---|---|---|---|
| MaterialCosts [sFr.] | $C_{Mat}$ | $C_{Mat} = C_{RAW} \cdot P_V(k)$ | $[0, \infty]$ | - | - |
| Turnover [sFr.] | $BV$ | $BV = P_{PU} \cdot P_V(k)$ | $[0, \infty]$ | - | - |
| RefusedOrderCosts [sFr.] | $C_{Refused}$ | $C_{Refused} = P_{RO} \cdot (P_{PU} - C_{RAW})$ | $[0, \infty]$ | - | - |

### Order Dispatch

| Influence Data | Symbol | Description | Codomain | Mean Value | Distribution | Data Origin |
|---|---|---|---|---|---|---|
| AverageRunTimeperUnitL [1/h] | $CT$ | Average production time for one unit | $[0, \infty]$ | 0.00217123 | - | Company |
| AverageIntermediateOrderEntryTime [h] | $T_{IT}$ | Time between two incoming orders | $[0, \infty]$ | 11.0866208 | Poisson (according to Lit. Res.) | Company |
| $\mu_{DemandQuantity}$ [pcs.] | - | Mean value of the demand quantity distribution | $[0, \infty]$ | 2'134.9 | Weibull | Company |
| $\sigma_{DemandQuantity}$ [pcs.] | - | Standard deviation of the demand quantity distribution | $[0, \infty]$ | 0.68444 | Weibull | Company |
| Setup Time | $T_{Setup}$ | Required setup time between two production orders | - | 0.2 | - | Company |

| Decision Variables | Symbol | Description | Codomain | Start Value | Step Size |
|---|---|---|---|---|---|
| Safety Factor [] | $s$ | Safety factor | $[1, 2]$ | 1 | 0.1 |

| Output Data | Symbol | Description | Codomain | Start Value | Distribution |
|---|---|---|---|---|---|
| Demand Distribution CV [] | $CV[D(t,T_i)]$ | Coefficient of variation of the demand distribution | $[0, \infty]$ | - | - |
| ProductionOrderDispatched [object] | - | The production order related information as $Q_{Order}$, $LT_j$ and $cLT_j$ are stored in this object | - | - | - |
| ProdRefusedOrder [pc.] | $P_{RO}$ | Sum of the volume of all refused orders | $[0, \infty]$ | - | - |

**Figure 5.17:** *Variable Definition for the Logistic Model*

## 5.2. Logistic Model

with

$$O_{Delayed}(k) : \text{Sum over all delayed orders}$$
$$P_O(k) : \text{Sum over all orders}$$

Values for output-variable cost for delayed products $C_{Delayed}$ can be calculated by:

$$C_{Delayed} = CR_{Delayed} \cdot O_{Delayed}(k) \qquad (5.18)$$

with

$$CR_{Delayed} : \text{Cost rate for delayed delivery}$$

Every time a finished order enters the *Logistic Model*, order quantity is determined and latched in the variable $Q_{Order}$. The value of this variable is used in section *Production Calculation*.

### 5.2.2 Production Calculation

Material costs $C_{Mat}$ and turnover $BV$ require the total amount of manufactured products $P_O(k)$ for evaluation.

$$C_{Mat} = C_{Raw} \cdot P_V(k)$$
$$BV = P_{PU} \cdot P_V(k)$$

with

$$C_{Raw} : \text{Raw material costs per unit}$$
$$P_{PU} : \text{Average revenue per unit}$$

Whenever an order is dispatched, the planned finishing time $T_{PFT}(i)$ of the order $i$ is compared with the latest finishing time of the already dispatched orders $T_{MPFT}$. The order is refused if:

$$T_{PFT}(i) < T_{MPFT}$$

The amount of all refused orders with their adjoint order quantities $P_{RO}$ is used to calculate the refused order costs $C_{Refused}$:

$$C_{Refused} = P_{RO} \cdot (P_{PU} - C_{Raw})$$

The values for $C_{Raw}$ and $P_{PU}$ in Figure 5.17 are modified due to industrial partner's regulations.

## 5.2.3 Order Dispatch

Since this model represents a Just-in-Time manufacturing of customized products, customer demand is directly converted into a production order (lot size = demand quantity). Hence, one customer demand results in a production order which is immediately dispatched at the time of arrival. See chapter "Maintenance and Logistics" for a detailed modelling description.

Demand, or consumption distribution, can be interpreted as an aggregation of several events during each period in time. Those events can be characterized by two independent distributions:

- Distribution of the frequency of events ($P(t, n)$ $\left[\frac{1}{h}\right]$)

- Distribution of the characteristic values of an event ($Q(t, z)$ $[pc.]$)

The combination of these two distributions results in demand distribution $D(t, T_i) = \int_{T_i} P(t, n) dn \cdot \int_{T_i} Q(t, z) dz$.

$$E[D(t, T_i)] = E[P(t, T_i)] \cdot E[Q(t, T_i)] \left[\frac{pc.}{h}\right] \quad (5.19)$$

$$CV^2[P^{D(t,T_i)}] = \frac{1 + CV^2[Q(t, T_i)]}{CV^2[P(t, T_i)]}$$

(5.20)

Amount of events per period in a pure stochastic process are Poisson-distributed with distribution function $P(n)$ [Schoensleben, 2002] and $E[P(t, T_i)] = \lambda_n \left[\frac{1}{h}\right]$. Incoming orders are typically described by such a pure stochastic process. The average interarrival time of incoming orders $T_{IT}[h]$ is:

$$T_{IT} = \frac{1}{E[P(t, T_i)]} \quad (5.21)$$

$$T_{IT} \sim \frac{1}{\mathcal{P}(\lambda_n)} = 11.087[h]$$

$$\lambda_n = 0.0902[h^{-1}]$$

## 5.2. Logistic Model

The values in the formula are derived from the demand distribution of the partner company. Let $T$ approach to 1, then interarrival time of incoming orders is $\frac{1}{\mathcal{P}(\lambda_n)}$ distributed.

Whereas literature provides reasonable advices for selecting an appropriate distribution for event frequency modeling, such guidelines are absent for order quantity estimate. For simulation, estimations about order quantity distribution are based upon historic data. Database of recorded order quantities incorporates 5000 data set entries and is used to identify a distribution that describes the data best. The act of finding the best distribution function is called distribution fitting. Distribution fitting is performed with the MLE (Maximum Likelihood Estimates) method. It is a statistical method of tuning the free parameters of a mathematical model (parameters of the probability distribution) to provide the best fit of the model to some reference data [Kay, 1993], [Lehmann and Casella, 1998]. After the fitting procedure the goodness of fit tests measure the compatibility of a random sample with a theoretical probability distribution function. In other words, these tests show how well the selected distribution fits to the data. Goodness of fit values are calculated with *EasyFit 4.0*, a statistic software that automatically computes the goodness of fit tests Kolmogorov-Smirnov, Anderson-Darling and Chi-Squared ( [Kececioglu, 1992]) and lists the proposed probability distribution according to their goodness of fit. Those analysis showed that demand quantity is best described with a two-parametric Weibull distribution ($\alpha = 0.68444, \beta = 2134.9$).

The probability-probability (P-P) plot 5.19 of the fitting shows inappropriate fitting accuracy. This plot is a graph of the empirical cumulative distribution function values plotted against the theoretical cumulative distribution function values of the model. If the specified theoretical distribution is the correct model this plot will be approximately linear and the graph of the model will coincide with the reference diagonal line. P-P plot and results from goodness of fit tests indicate that order quantity probability distribution is badly described by a Weibull distribution. Even a definition of lower and upper cut-off parameters to clear data set from outliers cannot significantly improve fitting accuracy. Nevertheless, a Weibull distribution for order quantity is assumed due to lack of better data.

Regarding equations in 5.19 expected demand $E[D(t, T_i)]$ per period can be estimated:

$$E[D(t, T_i)] = \lambda_n \cdot \alpha \cdot \Gamma\left(\frac{1}{\beta} + 1\right) \tag{5.22}$$

A normal distribution can be assumed when the coefficient of variation of the consumption distribution $CV[P^{D(t,T_i)}]$ [Schoensleben, 2002] is equal or lower than 0.4. The amount of orders in a period needed to expect a normal distribution $\lambda_{nmin}$ then follows

110                    Chapter 5. Maintenance, Production and Logistic Model

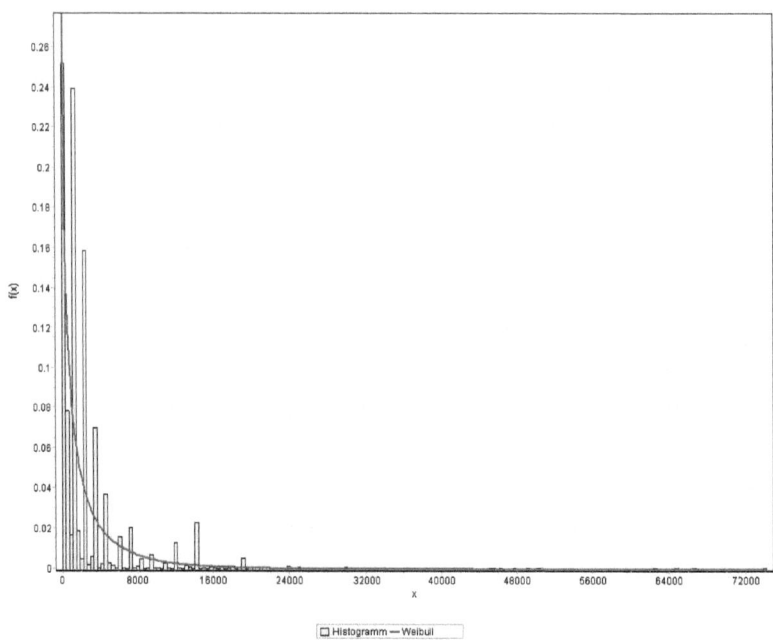

**Figure 5.18:** *Curve Fitting of Demand Quantity*

from the formula in 5.19. If $CV[Q(t,T_i)]$ is 1.7454, then at least 26 orders per period are necessary. The values are taken from the partner company.

$$CV^2[P^{D(t,T_i)}] = \frac{1 + CV^2[Q(t,T_i)]}{CV^2[n_{min}]}$$

$$0.4^2 = \frac{1 + 1.7454^2}{\lambda_{nmin}}$$

$$\lambda_{nmin} = \frac{4.046}{0.16} = 25.29$$

The minimal required sum of orders per period $\lambda_{nmin} \left[\frac{order}{h}\right]$ define a minimal planning horizon $T_{PH}[h]$:

$$T_{PH} \geq \frac{\lambda_{nmin}}{\lambda_n} \tag{5.23}$$

## 5.2. Logistic Model

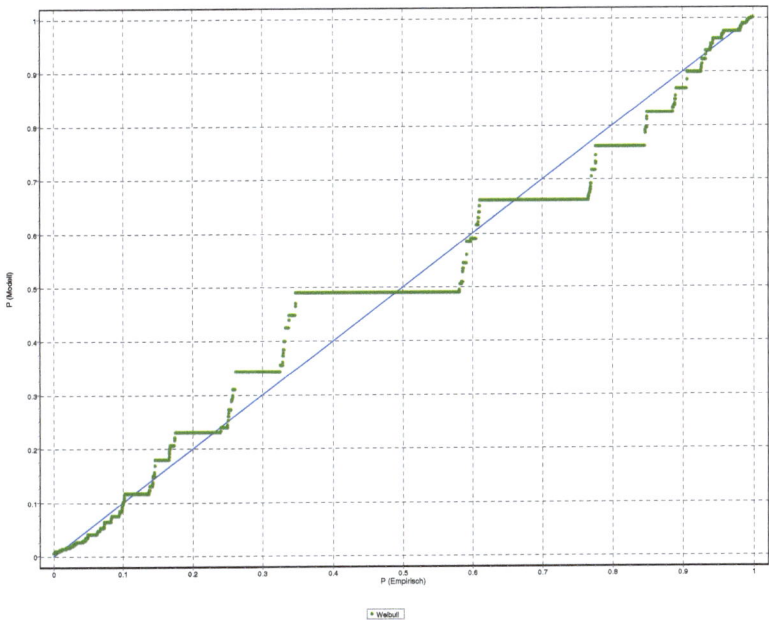

**Figure 5.19:** *Goodness of Curve Fitting*

Taking the values used in simulation, the minimum planning horizon is $\frac{25.29}{0.090199} = 280.38$ hours. Within this time-frame demand is considered to be *regular* or even *continuous* (following a normal distribution where outliers are very unlikely in contrast to heavy-tail distributions). Even demand is a basic prerequisite for reliable production planning and for simple control techniques such as Kanban [Halevi, 2001]. If the planning horizon is chosen too short, this quickly results in discontinuous and uneven demand patterns. Thus, requirement of smoothed demand sets a lower boundary for planning purposes. Since the planning horizon has no direct impact on other objective functions, its calculation can be computed offline.

### 5.2.3.1 Production Order

After an average time interval of $\frac{1}{E[P(t,T_i)]}$ a production order is released. In the model, a production order is realized as a message, or an object, with order specific information as order quantity $Q_{Order}[pc.]$, order finishing time $LT_j[h]$ and planned finishing time $cLT_j[h]$.

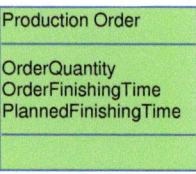

**Figure 5.20:** *Object "DispatchedProductionOrder"*

$Q_{Order}[pc.]$ : The amount of ordered products is recorded in this field
$LT_j[h]$ : Is the point in time when the order is finished and leaves the *Production Model*
$cLT_j[h]$ : Appoints the expected deadline
  $= (Q_{Order} \cdot CT + T_{Setup}) \cdot s$

with

$CT[h]$ : Average production time for one unit
$T_{Setup}[h]$ : Required setup time between two production orders

Estimation about the average production time per unit $CT$ is based on data provided by the partner company. Cycle times of the production system have been recorded for a time span of one year (1.3.2006 till 28.2.2007).

$$CT \approx 0.00166[h]$$

## 5.3 PRODUCTION MODEL

Dispatched production orders are processed in the production model (see Figure 5.21). It can be divided into a *Manufacturing* and *Production Calculation* section (P2.1 and P2.2 in Figure 5.21). Submodule *Manufacturing* represents the literal production process.

## 5.3. Production Model

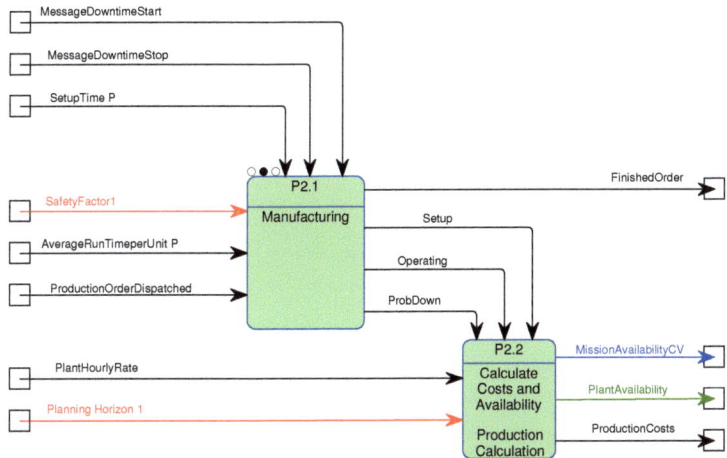

**Figure 5.21:** *Data Flow Diagram of the Production Model*

### 5.3.1 Manufacturing

In the model, production process is simplified depicted as a black box. Incoming orders are processed in the *Manufacturing* sub-modul and the finishing time of the production order is stored in the object "Production Order" of the flow "FinishedOrder". Residence times in the states "In Setup", "In Operating" and "In Down" are recorded and sent to the *Production Calculation* section.

Impact of maintenance on *Manufacturing* process is implemented is in the state chart 5.22, which incorporates the five states "In Idle", "In Setup", "In Operating", "In Down" and composite state "In Operating". An incoming order sets the state chart into state "In Setup", stays there for the time $T_{Setup}[h]$ and jumps into state "In Operation". Then, variable minimal production time $T_{MinProd}[h]$ is evaluated.

$$T_{MinProd} = Q_{Order} \cdot CT$$

Production process is set into operation. Seized order is released from the *Manufacturing* when the manufacturing process has been in state "Operating" for $T_{MinProd}$. The process starts again in state "In Idle".

Whenever a maintenance or repair activity is performed, a message "MessageDowntimeStart" is sent by the *Maintenance Model* and triggers the transition between state "In Operation" and "In Down" (compare with section 5.1). Present task is interrupted. When the maintenance or repair task is finished port "MessageDowntimeStop" is activated and the state chart is reset to the most recently visited state in the composite state "In Operating".

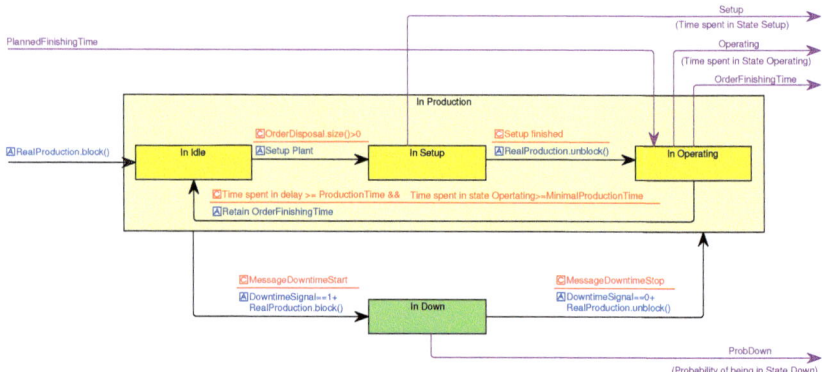

**Figure 5.22:** *State Chart Production Controlling*

### 5.3.2 Production Calculation

For computation of plant availability $A_{SS}(t)$ and production costs $C_P[CHF]$ accumulated residence time in the states "In Setup" (variable $D_{Setup}(t)[h]$) and "In Operating" (variable $D_{Operating}(t)[h]$) and the probability of being in state "In Down" ($P_{Down}(t)$) are required. Associated calculations are:

$$A_{SS}(t) = 1 - P_{Down}(t) \qquad (5.24)$$
$$C_P = (D_{Setup}(t) + D_{Operating}(t)) \cdot CR_{Plant} \qquad (5.25)$$

with

$$CR_{Plant} \; : \; \text{Cost rate of the plant} \left[ \frac{CHF.}{h} \right]$$

## 5.4. Top-Level Environment

Plant availability $A_{SS}(t)$ is a function of the chosen maintenance strategy and the adjoint failure rate which depend on time $t$. About the quality levels of the maintenance and repair activities, maintenance and failure rate devolution are interlinked and do mutually impact each other. Since the maintenance strategy and the failure rate are interdependent, an estimate about the plant availability progression can hardly be given and foster the application of simulation techniques.

Minimal planning horizon $T_{PH}[h]$ sets an additional constraint due to the requirement (compare with equation 3.31):

$$CV[P^{A_{Missions}(k,p,T_i)}] \leq 0.4$$

### 5.4 Top-Level Environment

*Main* is the environment in which the three submodules are embedded. On this level, global parameters and variables are defined to provide model-wide accessability to those constructs. Total costs $C_{Total}(t)[CHF.]$ and the Cash Flow $CF(t)[CHF.]$ are:

$$C_{Total}(t) = C_{Mat} + C_{Delayed} + C_P + C_{Rep} + C_{Prev} + C_{Refused}$$
$$CF(t) = BV - C_{Total}(t)$$

#### 5.4.1 Discounted Cash Flow

As described in subsection 1.1, the most profitable maintenance strategy is the one with the highest sum of discounted cash flows ($SumDCF(t)[sFr.]$). Calculation of the $SumDCF(t)$ provides a common basis for investment comparison. Thus, investments in sophisticated maintenance strategies have to compete with any other investment proposals.
For the $SumDCF(t)$ calculation the annual cash flows $CF_i(t)$ have to be determined first. Then, $DCF(t=j)[CHF.]$ of all cash flows $CF_i(t)$ up to year $j$ is:

$$SumDCF(t=j) = \sum_{i=1}^{j} \frac{CF_i(t)}{(1+p)^i} \quad (5.26)$$
$$CF_i(t) = BV_i - C_{Total,i}$$

## Production Model

### Manufacturing

| Influence Data | Symbol | Description | Codomain | Mean Value | Distribution | Data Origin |
|---|---|---|---|---|---|---|
| Setup Time | $T_{Setup}$ | Required setup time between two prodcution orders | - | 0.2 | - | Company |
| Average Run Time per Unit P [h/pc] | CT | Average production time for one unit | - | 0.00166 | - | Company |
| ProductionOrderDispatched [object] | - | The production order related information as $Q_{Order}$, $LT_j$ and $cLT_j$ are stored in this object | - | - | - | - |
| MessageDowntimeStop [] | - | Interrupts the production process during a repair or maintenance activity | [0, 1] | 0 | - | - |
| MessageDowntimeStart [] | - | Unblocks the production process after a repair or maintenance activity | [0, 1] | 0 | - | - |

| Decision Variables | Symbol | Description | Codomain | Start Value | Step Size |
|---|---|---|---|---|---|
| $s_T$ [] | - | Temporal safety factor | [1, 2] | 1 | 0.1 |

| Output Data | Symbol | Description | Codomain | Start Value | Step Size |
|---|---|---|---|---|---|
| FinishedOrder [object] | - | The production order related information as $Q_{Order}$, $LT_j$ and $cLT_j$ are stored in this object | - | - | - |
| Setup [h] | $D_{Setup}(t)$ | Time spent in state "In Setup" | [0, ∞] | - | - |
| Operating [h] | $D_{Operating}(t)$ | Time spent in state "In Operating" | [0, ∞] | - | - |
| ProbDown [] | $P_{Down}(t)$ | Probability of being in state "In Down" | [0, 1] | - | - |

### Production Calculation

| Influence Data | Symbol | Description | Codomain | Mean Value | Distribution | Data Origin |
|---|---|---|---|---|---|---|
| Plant Hourly Rate [sFr./h] | $CR_{Plant}$ | Cost rate of the plant | - | 180 | - | Company |
| Setup [h] | $D_{Setup}(t)$ | Time spent in state "In Setup" | [0, ∞] | - | - | - |
| Operating [h] | $D_{Operating}(t)$ | Time spent in state "In Operating" | [0, ∞] | - | - | - |
| ProbDown [] | $P_{Down}(t)$ | Probability of being in state "In Down" | [0, 1] | - | - | - |

| Output Data | Symbol | Description | Codomain | Start Value | Step Size |
|---|---|---|---|---|---|
| Mission Availability CV [] | $CV[A_{Mission,S}(k,p,T_i)]$ | Coefficient of variation of the mission availability distribution | [0, ∞] | 0 | - |
| PlantAvailability [] | $A_{SS}(t)$ | $A_{SS}(t) = 1 - P_{Down}$ | [0, 1] | 0 | - |
| ProductionCosts [sFr.] | $C_P$ | $C_P = (D_{Setup} + D_{Operating}) \cdot CR_{Plant}$ | [0, ∞] | 0 | - |

**Figure 5.23:** *Variable Definition for the Production Model*

with

$$BV_i \ : \ \text{Turnover in year } i$$
$$C_{Total,i} \ : \ \text{Total costs in year } i$$
$$p \ : \ \text{Interest rate}$$

## 5.4. Top-Level Environment

## Main

| Influence Data Main | Symbol | Description | Value |
|---|---|---|---|
| PreventiveMaintenanceCosts [sFr.] | $C_{Prev}$ | $C_{Prev} = (T_{PM}+T_{IM}+T_{MM}+T_{VM}+T_{SM}) \cdot k$ | |
| RepairCosts [sFr.] | $C_{Rep}$ | $C_{Rep} = (T_{PR}+T_{IR}+T_{MR}+T_{VR}+T_{SR}) \cdot k$ | |
| ProductionCosts [sFr.] | $C_P$ | $C_P = (D_{Setup}+D_{Operating}) \cdot CR_{Plant}$ | |
| DelayedDeliveryCosts [sFr.] | $C_{Delayed}$ | $C_{Delayed} = O_{Delayed}(k) \cdot CR_{Delayed}$ | |
| MaterialCosts [sFr.] | $C_{Mat}$ | $C_{Mat} = C_{RAW} \cdot P_V(k)$ | |
| RefusedOrderCosts [sFr.] | $C_{Refused}$ | $C_{Refused} = P_{RO} \cdot (P_{PU}-C_{RAW})$ | |
| Turnover [sFr.] | BV | $BV = P_{PU} \cdot P_V(k)$ | |
| Interest rate [] | p | Discount rate for discounted cash flow calculation | 0.15 |

| Output Data Main | Symbol | Description |
|---|---|---|
| CashFlow [sFr.] | CF(t) | $CF(t) = BV - C_{Total}(t)$ |
| TotalCosts [sFr.] | $C_{Total}(t)$ | $C_{Total} = C_{Mat}-C_{Delayed}-C_P-C_{Rep}-C_{Prev}-C_{Refused}$ |
| DiscountedCashFlow | SumDCF(t) | $DCF(t) = \sum CF_i/(1+p)^i$ |

**Figure 5.24:** *Variable Definition for the Main Section*

## Chapter 6

# Simulation and System Optimization

The model introduced in chapter 5 is implemented in a simulation environment called $AnyLogic^©$, a professional simulation engine to model and simulate complex hybrid, discrete and continuous discrete systems. Due to its object-oriented modelling approach, $AnyLogic^©$ offers an ease-of-use possibility to represent a system on different levels of details and provides reusability of simulation modules.

This chapter discusses results of numerous simulation runs of the model created in the previous chapter. Among the investigation of the effect of preventive maintenance on system availability and cost effectiveness, different maintenance strategies shall be optimized with respect to maximize system availability $A_{SS}(t)$, cash flow $CF(t)$ and discounted cash flow $SumDCF(t)$.

### 6.1 GOALS OF THE SIMULATION

Aim of the simulation is to investigate the availability $A_{SS}(t)$, the accumulated cash flow $CF(t)$, the sum of discounted cash flows $SumDCF(t)$ and the service level $S_L(t)$ of the production system with parameters as introduced. Then, the impact of the safety factor $s$ on $A_{SS}(t)$, $CF(t)$ and $SumDCF(t)$ shall be analyzed and the effect of intensified preventive maintenance on $CV[P^{A_{Missions}(k,p,T_i)}]$ is to be studied. The three simulation experiments are defined as following:

1. Default production system with the introduced parameters (Exp. 1)

2. Default production system without preventive maintenance (Exp. 2)

3. Investigating the effect of a step-wise increased safety factor $s$ on production system availability and cash flow (Exp. 3)

## 6.2 DESIGN OF EXPERIMENTS

An experiment plan is used to organize the execution of different simulation runs under varied start conditions and parameter values in a systematic order. To meet the statistic requirements on quantity, quality, and accuracy of the simulation results under the condition that not any possible combination of parameter variations were considered, a statistic experiment planning is requested.

Sensitivity analysis, factorial test planning, partial factorial test planning, Taguchi- and Shainin method are well established techniques in statistic planning of simulation experiments. Those methods are aimed at reducing model complexity by identifying correlations and interdependencies between different parameters and variables (input-output interactions). Weak interactions can be neglected and their related input and output values, eventually, be excluded from the simulation model.

**Sensitivity analysis** In a sensitivity analysis only one parameter is changed at the same time while all others remain constant. The effect is an influence data separation, while the impact of combinations of influence data remains unconsidered. This method is suitable to investigate single and independent influence data.

**Factorial and partial factorial test planning** All influence data are changed simultaneously so that interdependencies can be observed. Hence, interdependency does not describe the impact of a single influence data on another but the combined effect of several influence data on the target value. If all possible influence parameter combinations are considered, it is called a factorial experiment and allows a complete analysis of single influence data and all interplays. On the contrary, a partial factorial experiment incorporates only a limited amount of combinations of influence data. Partial factorial test planning is less time-consuming but also less accurate in terms of identifying interdependencies in influence data than factorial test planning.

**Taguchi and Shainin Method** Both methods originate from the statistic test planning in quality assurance. Taguchi assumes that most interdependencies can be neglected, whereas Shainin states that a problem, that eventually can have various influence data, is reducible to only a few causes (Pareto-Principle).

## 6.2. Design of Experiments

In this simulation, sensitivity analysis is used to study the impact of the safety factor on service level $S_L(t,k)$, cash flow $CF(t)$ and discounted cash flow $SumDCF(t)$ (Exp. 3). Moreover, estimates about start conditions, step sizes and the variation range of the decision parameters in the optimization experiments are elaborated with sensitivity analysis (see Figure 6.8). Within this concept of simulation experiment, some terms have to be defined and clarified. They follow the definitions in [VDI-3633-Blatt-1, 1993].

**Definition 26** *Target values are the expected results of an experiment. They depend on the values of the decision variables. In the case of optimization, target values constitute the objective function (the optimization criteria).*

**Definition 27** *A decision variable is an unknown quantity representing a decision that needs to be made. This variable is controlled by the decision maker and constitutes the impact the decision maker has on the system.*

Those variables define the maintenance strategy in the model in terms of frequency and quality level (combination of $I_{PM}$, $I_{IM}$, $I_{MM}$, $P_{ImperfectR}$, $P_{MinimalR}$ and $s$). Those decision variables have to be chosen in accordance with the optimization criteria and are printed in green in the figure 5.14. The optimizer varies decision variables in search of values that maximize or minimize the objective function.

**Definition 28** *Step size depicts the minimal possibility of variation of the decision variables.*

A definition of step sizes requires a discretisation of the decision variables. This discretisation is often a difficult task and is always a trade-off between objective accuracy and simulation duration.

**Definition 29** *Variation range limits the possibilities of variation with an upper and lower boundary.*

Elaboration of the variation range exposes the same problems as the definition of the step sizes.

## 6.3 Simulation Run of the Default Production System (Exp. 1)

In the first section, results of the default system are presented; "default" refers to unchanged influence parameters as defined in Figure 6.1.

### Simulation

| Duration Simulation Run | $T_{maxPH}$ | | 64'000 [h] |
|---|---|---|---|
| Measuring Intervals | | | 1 [h] |

| Target Values | Symbol | Description | |
|---|---|---|---|
| Availability [] | $A_{SS}(t)$ | | |
| CashFlow [sFr.] | CF(t) | CF(t) = BV-$C_{Total}$(t) | |
| DiscountedCashFlow [sFr.] | SumDCF(t) | DCF(t)=∑CF/(1+p)^i | |
| ServiceLevel [] | $S_L$(t,k) | $S_L$(t,k) = 1-$O_{Delayed}$(k)/$P_O$(k) | |

| Decision Variables | Symbol | Description | Start Value |
|---|---|---|---|
| PerfectMInterval [h] | $I_{PM}$ | Time between two perfect preventive maintenance activities | - |
| ImperfectMInterval [h] | $I_{IM}$ | Time between two imperfect preventive maintenance activities | 100 |
| MinimalMInterval [h] | $I_{MM}$ | Time between two minimal preventive maintenance activities | - |
| PImperfectR [] | $P_{ImerfectR}$ | Probability indicator for imperfect repair | 1 |
| PMinimalR [] | $P_{MinimalR}$ | Probability indicator for minimal repair | 0 |
| Safety Factor [] | s | Safety factor | 1 |

**Figure 6.1:** *Simulation Start Conditions*

### 6.3.1 Model Accuracy

The model is tested against real data recorded between the 1.3.2006 till 28.2.2007. As reference parameter the mission availability $A_{Mission}(t, T_i)$ with $T_i = 413.2[h]$ is taken since only for this parameter more or less reliable data exist. Those values are compared with the simulated average monthly system availabilities (see Figure 6.2)

## 6.3. Simulation Run of the Default Production System (Exp. 1)

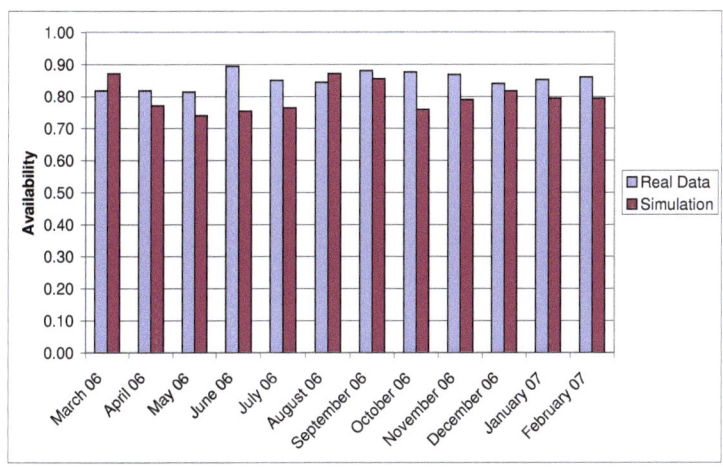

**Figure 6.2:** *Comparison of Monthly System Availability*

The coefficient of correlation $R$ is an indicator for the model accuracy in comparison to real data (see [Papula, 1999] for further explanations). $R$ is defined on the co-domain of $[-1, 1]$ whereas the three boundaries -1, 0 and 1 represent a negative, no, or a positive coherence between the simulation and the effective results. In case of high model accuracy, the $R$-value close to 1; all points should be on one line (compare with Figure 6.3).

The coefficient of correlation of the given model is fairly good ($R = 0.764$).

### 6.3.2 Availability and Service Level

Availability curve (printed in red) in Figure 6.4 shows the characteristic decreasing profile caused by increasing failure rates and immediate impact of preventive and corrective maintenance. With increasing simulation time, availability exposes a decreasing trend. It is not apparent if availability ever stabilizes and the system merges into a steady-state, or that deterioration of the system in significantly mitigated. Simulation runs over longer simulation times ($t = 10^6$ hours) confirmed decelerating availability in time and the absence of any stabilization. System cannot reach a steady-state since $\lambda_{Failure}(t)$ increases in time and this deteriorating cannot be stopped with preventive maintenance (non-maintainable part of the failure rate).

124                              Chapter 6. Simulation and System Optimization

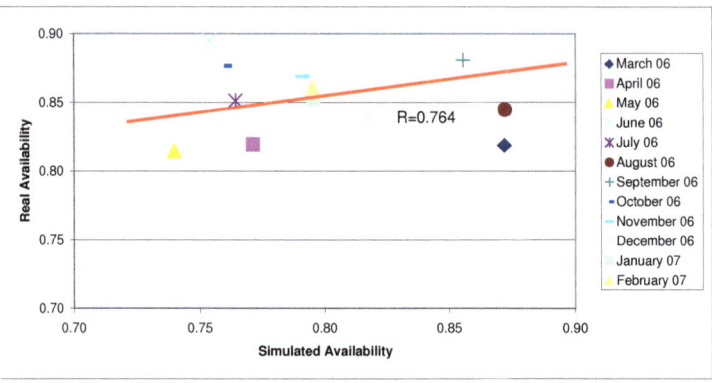

**Figure 6.3:** *Scattergram of Real and Simulated Availability Values*

The qualitative curve progression of the service level follows the availability graph but on a lower value level and with a temporal delay. Since the service level is a function of availability, this is not surprising. Moreover, the service level decreases faster and stronger than the system availability. Their final values at the end of the simulation time are depicted in table 6.2.

The low end value of the service level in Figure 6.5 can be explained by the low safety factor $s = 1$. Regarding the results of the mission availability distribution (see table 6.3), a planning horizon of $T_{PH} = 1'000$ obeys the requirement of $CV[P^{A_{Missions}}(k,p,T_i)] \leq 0.4$. $CV[P^{A_{Missions}}(k,p,T_i)]$ is calculated on the basis of the histogram with the amount of classes $g$. This amount of classes is predicated on the recommendation ( [Vogel, 2003]) that:

$$g \leq 5 \cdot log(n)$$

with

$$n \quad : \quad \text{Sum of data}$$

Since $n = 64$ for $T_{PH} = 1'000$, sum of classes is 9. The histogram of the system availability exposes a strong concentration of values in the range between 0.8182 and 1 and is the reason for the low coefficient of variance of the distribution.

A simulation run with $T_{PH} = 280$ is performed to check if $CV[P^{A_{Missions}}(k,p,T_{PH}=280)]$ obeys the condition of $CV[P^{A_{Missions}}(k,p,T_{PH}=280)] \leq 0.4$. Associated value for the coef-

## 6.3. Simulation Run of the Default Production System (Exp. 1)

|           | Real Data | Simulation Results |
|-----------|-----------|--------------------|
| March     | 0.82      | 0.87               |
| April     | 0.82      | 0.77               |
| May       | 0.81      | 0.74               |
| June      | 0.89      | 0.75               |
| July      | 0.85      | 0.86               |
| August    | 0.84      | 0.87               |
| September | 0.88      | 0.86               |
| October   | 0.88      | 0.76               |
| November  | 0.87      | 0.79               |
| December  | 0.84      | 0.82               |
| January   | 0.85      | 0.80               |
| February  | 0.86      | 0.80               |
| Variance  | 0.00068   | 0.00178            |
| Mean      | 0.85      | 0.80               |

**Table 6.1:** *Comparison of Simulation and Real Data*

| $A_{SS}(t = 64'000)$     | 0.882 |
|--------------------------|-------|
| $S_L(t = 64'000, k)$     | 0.005 |

**Table 6.2:** *Simulation Results of $A_{SS}(t = 64'000)$ and $S_L(t = 64'000, k)$*

ficient of variance of $P^{A_{Missions}(k,p,T_{PH}=280)}$ is equal to 0.063 and far below the critical boundary of 0.4. Thus, the minimal planning horizon is determined by the demand distribution.

| $CF(t)$                               | $4.20 \cdot 10^7$ |
|---------------------------------------|-------------------|
| $SumDCF(t)$                           | $2.10 \cdot 10^7$ |
| $E[P^{A_{Missions}(k,p,T_i)}]$        | 0.880             |
| $\sigma(P^{A_{Missions}(k,p,T_i)})$   | 0.055             |
| $CV[P^{A_{Missions}(k,p,T_i)}]$       | 0.063             |

**Table 6.3:** *Simulation Results for $P^{A_{Missions}(k,p,T_i)}$ of Experiment 1*

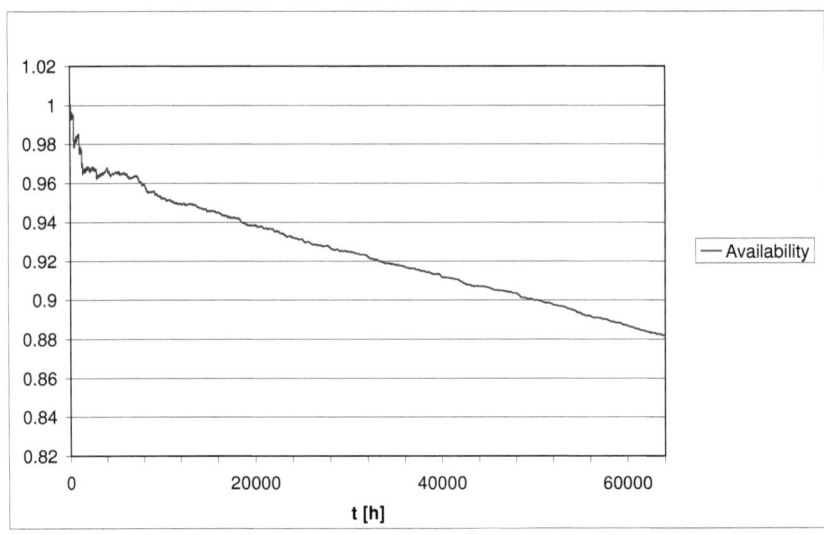

**Figure 6.4:** *Progression of $A_{SS}(t)$*

## 6.4 PRODUCTION SYSTEM WITHOUT PREVENTIVE MAINTENANCE (EXP. 2)

For this purpose, the decision variables $I_{PM}$, $I_{IM}$ and $I_{MM}$ are set to zero to avoid preventive maintenance activities. All other parameters are left unchanged.
Those simulation results are compared with the preventive maintained production sys-

|  | System with PM | System without PM |
|---|---|---|
| $CF(t)$ | $4.20 \cdot 10^7$ | $4.18 \cdot 10^7$ |
| $SumDCF(t)$ | $2.10 \cdot 10^7$ | $2.09 \cdot 10^7$ |
| $S_L(t,k)$ | 0.005 | 0.006 |
| $A_{SS}(t)$ | 0.882 | 0.909 |
| $CV[P^{A_{Mission_S}(k,p,T_i)}]$ | 0.063 | 0.070 |

**Table 6.4:** *Comparison between Preventive and Corrective Maintained Production Systems*

## 6.5. Sensitivity Analysis of the Safety Factor (Exp. 3)

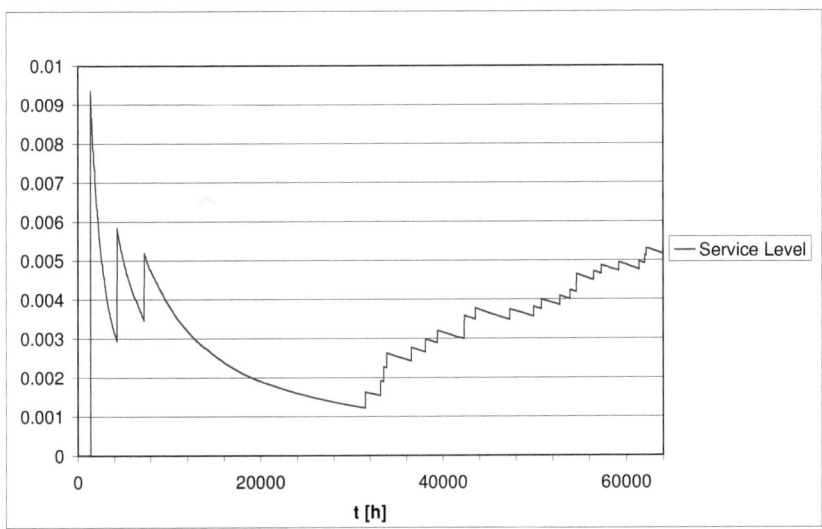

**Figure 6.5:** *Progression of $S_L(t)$*

tem (please consult table 6.4).

The values in table 6.4 indicate that the given system tends to be slightly over-maintained with regards to the achieved system availability. Furthermore, a comparison of the $CV[P^{A_{Mission_S}(k,p,T_i)}]$ between the preventive and corrective maintained system corroborates the believe that preventive maintenance has a positive impact on the uncertainty in the system (10% reduction of $CV[P^{A_{Mission_S}(k,p,T_i)}]$).

## 6.5 Sensitivity Analysis of the Safety Factor (Exp. 3)

Effect of an increased safety factor $s$ from 1 to 100 on the objective values $A_{SS}(t)$ (compare with definition 1.4.1), $CF(t)$ (see definition 1.3 ), $SumDCF(t)$ (equation 1.1) and $S_L(t,k)$ (following the definition in 1.4.6) is studied. Results, shown in table 6.5, highlight the impact of the safety factor not only on the logistics parameters but also on system availability $A_{SS}(t)$.

| $s$ | 1 | 2 | 3 | 4 | 5 |
|---|---|---|---|---|---|
| $CF(t)$ | $4.20 \cdot 10^7$ | $3.50 \cdot 10^7$ | $2.56 \cdot 10^7$ | $1.86 \cdot 10^7$ | $1.12 \cdot 10^7$ |
| $SumDCF(t)$ | $2.10 \cdot 10^7$ | $1.76 \cdot 10^7$ | $1.28 \cdot 10^7$ | $9.21 \cdot 10^6$ | $5.98 \cdot 10^6$ |
| $S_L(t,k)$ | 0.005 | 0.910 | 0.951 | 0.974 | 0.987 |
| $A_{SS}(t)$ | 0.884 | 0.881 | 0.877 | 0.881 | 0.879 |

**Table 6.5:** Impact of $s$ on $CF(t)$, $SumDCF(t)$, $S_L(t,k)$ and $A_{SS}(t)$

## 6.5.1 Impact of the Safety Factor on System Availability

The stringent relation between safety factor and system availability, proposed in section 3.3.1, can be verified with the simulation.

The safety factor $s$ causes a stretching of the seizing time of the production system per order. This stretching effect may have a positive impact on system availability since periods of high system availability may coincide with the seizing time, respectively the production system fails in idle times. Safety factor impacts synchronization of seizing time and high production system availability. However, although a connection between safety factor and system availability can be proven, an analytical estimation of the impact of a deviated safety factor can hardly be given due to system complexity.

## 6.5.2 Effect of the Safety Factor on the Service Level

An increasing safety factor has a dominant impact on the service level. This is not amazing since it is assumed that the service level is mainly controlled by the chosen safety factor (compare with equation 3.26). Regarding Figure 6.6, the increase of the safety factor from 1 to 2 causes the steepest rise of the service level, and the impact of the safety factor decreases with rising value of the safety factor and degenerates to zero when the service level reaches 1. However, a large safety factor causes that many orders are rejected and the production system is under-utilized and unprofitably operated (see subsection 6.5.3). Thus, taking the service level as only objective parameter without respecting its financial aspects is inadequate to optimize the overall performance of a production system.

## 6.6. Optimization

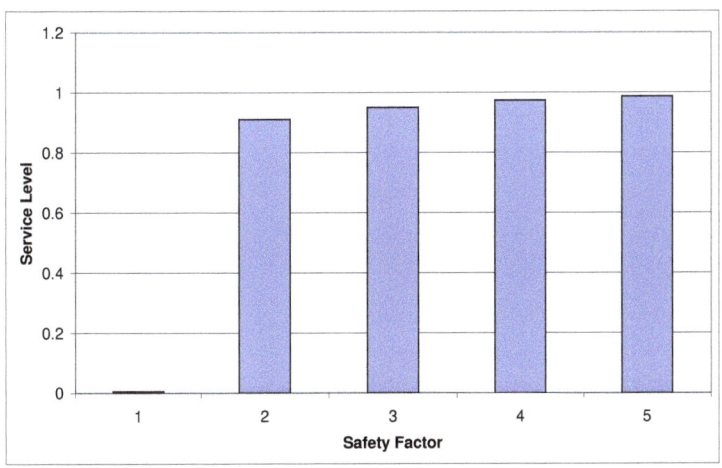

**Figure 6.6:** *Impact of different Safety Factors on Service Level*

### 6.5.3 Influence of the Safety Factor on the Cash Flow and Discounted Cash Flow

Variation of the safety factor has a strong influence on the cash flow (see Figure 6.7). A high safety factor causes that a large amount of orders are rejected and the production system exposes an inadequate utilization. Rejected orders account for opportunity costs since they represent loss of profit and those costs minimize the cash flow. Moreover, a low utilization brings about a low system output and revenue. Hence, a safety factor chosen too high has a negative impact on the cash flow and the overall system profitability.

The bars in figure 6.7 expose a clear trend towards decreasing $CF(t)$ and $SumDCF(t)$ with increasing safety factor $s$.

### 6.6 OPTIMIZATION

Importance of simulation optimization is that most real world problems in optimization are too complex to be described in mathematical formulations. Nonlinearities, combinatorial relationships or uncertainties often give rise to simulation as the only possible so-

# Chapter 6. Simulation and System Optimization

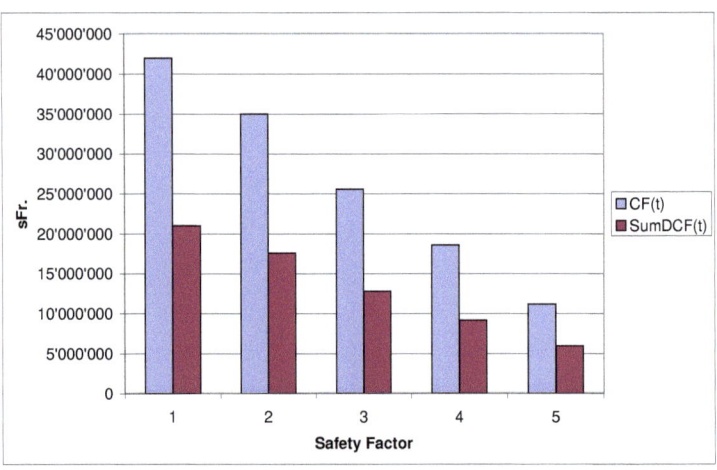

**Figure 6.7:** *Impact of different Safety Factors on $CF(t)$ and $SumDCF(t)$*

lution approach. Since classical optimization methods, as linear, non-linear and mixed integer programming fail, recourse is made to a "scenario generator" that at least one scenario will provide an admissible solution. Creation of this "scenario generator" to produce acceptable and high quality solutions has a long history in the simulation and optimization community.

Some heuristic approaches for optimizing maintenance strategies can be found in appendix D.

## 6.6.1 Optimizing Techniques

"Scenario generator" refers to the task of creating a way to guide a series of simulations to produce solutions that converge as quick as possible to the optimal solution.

Fu [Fu, 2002] identified four main approaches for optimizing simulations:

**Stochastic approximation (gradient-based approaches)** This method imitates the gradient search method used in deterministic optimization. According to the estimate of the objective function gradient, the search direction is determined. When the gradient of the objective function is evaluated to zero, optimization algorithm ter-

## 6.6. Optimization

Optimization

| Duration Simulation Run | $T_{maxPH}$ | | 64'000 [h] | approx. 10 years |
| Measuring Intervals | | | 1 [h] | |

| Target Values | Symbol | Description | Codomain | Optimization |
|---|---|---|---|---|
| Availability [] | $A_{SS}(t)$ | | [0, 1] | Max |
| TotalAmountOfCoverage [sFr.] | AoC | AoC = BV-$C_{Total}$ | [-∞, ∞] | Max |
| DiscountedCashFlow [sFr.] | SumDCF(t) | DCF(t)=$\sum CF_i/(1+p)^i$ | [-∞, ∞] | Max |
| ServiceLevel [] | $S_L(t,k)$ | $S_L(t,k) = 1-O_{Delayed}(k)/P_O(k)$ | [0, 1] | Max |

| Decision Variables | Symbol | Description | Codomain | Start Value | Step Size |
|---|---|---|---|---|---|
| PerfectMInterval [h] | $I_{PM}$ | Time between two perfect preventive maintenance activities | [1'000, 64'000] | 50'000 | 1'000 |
| ImperfectMInterval [h] | $I_{IM}$ | Time between two imperfect preventive maintenance activities | [100, 64'000] | 5'000 | 100 |
| MinimalMInterval [h] | $I_{MM}$ | Time between two minimal preventive maintenance activities | [100, 64'000] | 1'000 | 100 |
| PImperfectR [] | $P_{ImperfectR}$ | Probability indicator for imperfect repair | [$P_{MinimalR}$, 1] | 1 | -0.1 |
| PMinimalR [] | $P_{MinimalR}$ | Probability indicator for minimal repair | [0.01, $P_{ImperfectR}$] | 0.71 | 0.1 |
| Safety Factor [] | s | Safety factor | [1,5] | 1.5 | 0.1 |

| Optimization Stop Conditions | |
|---|---|
| Maximum Number of Simulations | 10000 |
| Objective Function Precision | 0.01 |

**Figure 6.8:** *Optimization Plan*

minates [Gerencsér, 1999].

**(Sequential) response surface methodology** A response surface is created by tracking the response achieved from running the simulation with different input values. This response surface can be used as simplified meta-model, which locally approximates the response surface, for optimizing the original model. The "local" meta-model is applied to define a search strategy (e.g., step forward in the estimated gradient direction) and this procedure is repeated [Montgomery and Myers, 2002].

**Random search** This techniques randomly selects a point from the neighborhood of the current point and requires a definition of the neighborhood in the search algorithm. Its attractiveness derives from the characteristics of almost sure convergence under very general conditions [Spall, 2003]. However, convergence alone is an insufficient indicator for efficiency of the algorithm. Rate of convergence is also of interest. For multi-dimensional search space the method is a very slow algorithm since search space exponentially growths as the sum of decision variables increases.

**Sample path optimization** Principle idea is to optimize a deterministic function that bases on limited sum of simulation runs $n$ [Guerkan et al., 1994]. Those simulation runs are initiated with different values of the input factors and are averaged to derive an approximation of the true characteristic of the simulation. Optimization is performed on this deterministic function. To provide sufficient accuracy of the approximation, $n$ needs to be large [Andradóttir, 1998].

Although these four approaches are widely spread in the literature, they have hardly been implemented in optimization for simulation software due to their complexity and computational power consumption [April et al., 2003]. Application of those methods require a considerable understanding about their technique on the part of the user. Those deficiencies and the exponential increase of computational power promote application of metaheuristic approaches.

Metaheuristic approaches consider the simulation model as black box [April et al., 2003]. An optimizer assigns rule-based chosen values to the decision variables and uses the simulation response to decide about the selection of values for the next run. Decisions in the optimizer are based on e.g. evolutionary algorithms or simulated annealing that combines two or more solutions out of a population for the next run. The population contains a selection out of all previous simulation results.

Principle advantage of evolutionary approaches over those described above is their capability to investigate a larger area of the solution space with less simulation runs. $AnyLogic^{©}$ provides an optimization engine $OptQuest^{©}$. $OptQuest^{©}$ uses methods that integrate state-of-the-art metaheuristic procedures, including tabu search, neural networks, and scatter search, into a single composite method.

### 6.6.1.1 Scatter Search in $OptQuest^{©}$

Scatter search combines composite decision rules and surrogate constraints. It differs by joining solutions and using strategic designs where other approaches, such as genetic algorithms, are limited to randomized search [Laguna, 2002]. The optimization methodology is based on the premise that problem solving must incorporate some kind of memory. Thus, earlier feasible solutions are taken into consideration for creating new solution approaches. Scatter search reconfigures new solutions out of a pool of solutions, the reference set. This new solutions are created by a convex or non-convex combination of two or more solutions and added to the pool of solutions if they have membership in the "best" solution. "Best" is not only referred to the value given by the objective function but also a measurement of increasing the diversity of the reference

## 6.6. Optimization

set and is based on predefined rules and a neural network. This neural network is used to represent a metamodel of the problem. With help of this metamodel, potentially bad solutions can be filtered out before they get evaluated on the real model. Thus, applying a metamodel increases the performance of the optimization process since only solutions that have passed the neural network filter are evaluated with the real model. The predefined rules for deciding whether the solution has membership in the "best" solutions or not are an offspring of the tabu search concept and are related with its four dimensional memory structure [Glover and Laguna, 2002].
The scatter search approach may be outlined as:

**Primary Generation Creation** to generate a collection of start solutions with the goal of creating a diversified population. The measurand of diversification is based on the Euclidian distance measure, which defines how "close" a potential new solution is from the solutions already in the population [Laguna, 1997].

**Feasibility Test** to enhance the performance of the optimization by preventing the optimization from simulation runs with bad solutions. This pre-processed filtering is either done with approximating the optimization task as a mixed-integer programming problem or with a neural network approach.

**Subset Creation Method** is used to generate a selection of solutions as basis for further combined solutions. Two reference points out of the population are selected to create four linear-combinatoric offsprings. Let $X_1$ and $X_2$ be the beginning reference points, then the offsprings $X_3$ to $X_6$ are:

$$\begin{aligned} X_3 &= X_1 + d \\ X_4 &= X_1 - d \\ X_5 &= X_2 + d \\ X_6 &= X_2 - d \end{aligned}$$

where $d = \frac{X_1 - X_2}{3}$. The selection of the parent solutions is subject to the measure of attractiveness. Attractiveness is defined as a combination of the age of the solution in the reference set and its objective function value. This set of new solutions is undergone the feasibility test and the reference membership method.

**Reference Set Membership Method** is required to create and enhance a reference set containing the "best" solutions. This method is based on the four dimensional memory structure introduced in the tabu search concept.

**Determination Conditions** defines the termination of the optimization process, either depending on the accuracy of the objective function values or on the sum of optimization runs.

Regarding local optimality, $OptQuest^©$ includes a number of metaheuristic solvers providing that a local optima can be left. The main solver is based on the scatter search methodology. Other solvers are based on popular procedures such as genetic algorithms and particle swarm optimization. Unless specified by the user, $OptQuest^©$ defaults to the scatter search solver [Laguna, 2008].

$OptQuest^©$ does provide functionality for bi-objective optimization. Bi-objective optimization means that two objectives can be optimized simultaneously. However, this feature is linked to risk analysis problems and is not integrative part of the optimization engine implemented in $AnyLogic^©$. Thus, the solver provides one-dimensional optimization only.

### 6.6.2 Optimization Stop Conditions

The optimization technique "Scatter Search" is aimed at rapidly converging to the global maximum or minimum of the optimization criteria. However, terminating criteria are needed to decided whether the value of the objective function is significant or further optimization runs are required. Without those additional criteria the optimizer is urged to compute all possible combinations of the decision variables which is extremely time consuming. On an Intel (R) Xeon 3.00 GHz server with 2 GB RAM, a single optimization run of the given example takes around 38 seconds, and there exist approximately $1.04 \cdot 10^{10}$ possible combinations of decision variable values. This results in a maximum required computing time of over 12'000 years per optimization. Thus, optimization stop conditions are introduced that interrupt optimization if one of the two conditions is satisfied:

- Maximum Number of Simulations is exceeded
- Value of the Objective Function stops improving (Automatic Stop)

Automatic stop corresponds to the situation when the objective function stops improving, meaning that its values differ less than $p$ during $n$ iterations. $p$ is the objective function precision and $n$ is calculated as 5% of the maximum number of simulations. The number of simulations determines both, stop condition and the search strategy of the $OptQuest^©$ engine. With decreasing number of simulations, the optimization engine uses a more aggressive search strategy.

## 6.6.3 Optimization Criteria

Objective functions are the target functions in focus on which the optimization criteria are deployed, either minimizing or maximizing the objective function by varying the decision variables. An objective function is a mathematical representation of the optimizing criteria.

Optimization process consists of repetitive simulation runs with different values of the influence variables. Those variables are varied from simulation to simulation to find the optimal combination of parameter values to solve the problem with respect to the objective function and constraints (compare with Figure 6.8).

Although a multitude of different optimization criteria exist in maintenance strategies, they are commonly addressed to meet one or a combination of the following, concurring goals [Warnecke, 1992]:

1. Minimal Overall Costs

2. Maximum System Availability

These are the classic criteria in the field of maintenance strategy optimization. Besides of those optimization criteria, a preferential maintenance strategy should meet some additional constraints, mostly covering safety and production aspects.

Besides of these classic optimization criteria, the proposed model offers additional objective functions due to the model's extensions towards logistic purposes.

3. Maximum Service Level

4. Maximum Cash Flow

5. Maximum Discounted Cash Flow

However, minimizing the overall costs is rather unprofitable without incorporating the revenue under those optimized conditions. Costs have to be contrasted with the revenue under those conditions to estimate the profitability of this setting. Thus, it is more useful to maximize the difference between revenue and costs - the cash flow.

Regarding the setup of the model, service level rises with increasing safety factor. Since service level is mainly impacted by the chosen safety factor, a maximizing of the service level would call for an increase of the safety factor until the service level reaches the overall maximum of 1. Moreover, an isolated consideration of the service level without respecting the financial aspects is meaningless since a high service level causes many rejected orders and a low production system utilization. Thus, only three of the former five optimization objectives turn up to be reasonable:

**Opt. 1** Maximum System Availability

**Opt. 2** Maximum Cash Flow

**Opt. 3** Maximum Discounted Cash Flow

Optimization task is to find an optimal, with respect to the optimization criterion, combination of different maintenance tasks, maintenance intervals and quality levels which do not violate additional constraints. Additional constraints come from production planning and controlling as a maximum amount of unplanned interruptions, fix maintenance intervals or job safety requirements. The only basic conditions in the model are:

$$T_{PH} \ni T_i \begin{cases} CV[P^{A_{Mission_S}(k,p,T_i)}] \leq 0.4 \\ CV[P^{D(t,T_i)}] \leq 0.4 \end{cases}$$

Since those constrains are limited to planning reasons and the planning procedure is not part of the simulation, those constrains are not integrated in this optimization.

The problem with optimizing for a single objective function reveals the problem that it often ends in wrong conclusions. Optimizing for another objective function, such as system availability or cash flow, might yield a different optimal maintenance policy. This deficiency could be overcome by defining minimal requirements for the other objective functions and introducing them as additional constraints. However, any additional constraint truncates the solution space that eventually can end up in finding no feasible solution.

The choice of appropriate maintenance actions is represented by the maintenance intervals of the corresponding maintenance tasks. Those predefined maintenance intervals are some of the decision variables.

### 6.6.4 Maximizing $A_{SS}(t)$ (Opt. 1)

Optimization of system availability $A_{SS}(t)$ terminated after 504 optimization runs and lead to a drastic reduction of preventive maintenance activities in comparison with the default system (see table 6.6 and Figure 6.9). Whenever financial or profitability aspects

## 6.6. Optimization

are in the focus of optimization, system availability cannot be taken as objective function. Without considering other measurands as $CF(t)$ or $SumCF(t)$, maximizing $A_{SS}(t)$ will not optimize the financial facets of the problem. This spotlights the deficiency of optimizing a single objective function. The minimal planning horizon $T_{PH}$ is not affected by the mission availability distribution $P^{A_{Missions}(k,p,T_i)}$ but is determined by the demand distribution $P^{D(t,T_i)}$.

### 6.6.5 Maximizing $CF(t)$ (Opt. 2)

The simulation experiment of maximizing the cash flow is addressed to maximize the profitability of the production system. Unless in subsection 1.3.1, $CF(t)$ is the cash flow over time without respecting the time value of money. The results are given in table 6.7 and the corresponding optimal maintenance strategy is given in Figure 6.10. Maximizing $CF(t)$ causes a drift towards later maintenance activities in comparison to any other optimization experiments. Optimization terminated after 1111 optimization runs and provides the following results:

### 6.6.6 Maximizing $SumDCF(t)$ (Opt. 3)

The dynamic approach incorporates the time value of money and attempts to maximize $SumDCF(t)$. Within this concept of time value of money, the chosen discount rate is of significant importance (compare with subsection 1.1).

| | |
|---|---|
| $I_{PM}$ | 64'000 |
| $I_{IM}$ | 63'400 |
| $I_{MM}$ | 58'200 |
| $P_{ImperfectR}$ | 1.0 |
| $P_{MinimalR}$ | 0.41 |
| $s$ | 3.5 |
| $CF(t)$ | $2.16 \cdot 10^7$ |
| $SumDCF(t)$ | $1.10 \cdot 10^7$ |
| $S_L(t,k)$ | 0.978 |
| $A_{SS}(t)$ | 0.913 |
| $T_{PH}$ | 280.38 |

**Table 6.6:** *Optimal Values of the Decision Variables to maximize $A_{SS}(t)$*

138    Chapter 6. Simulation and System Optimization

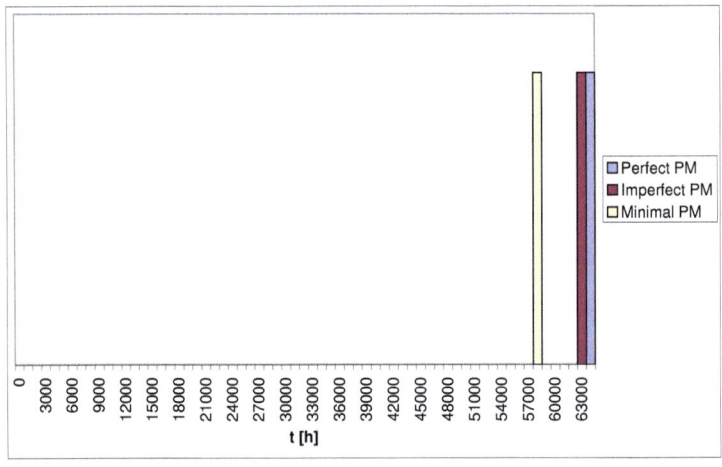

**Figure 6.9:** *Optimal Maintenance Strategy for Maximizing $A_{SS}(t)$*

| | |
|---|---|
| $I_{PM}$ | 43'000 |
| $I_{IM}$ | 41'300 |
| $I_{MM}$ | 52'200 |
| $P_{ImperfectR}$ | 1.00 |
| $P_{MinimalR}$ | 0.11 |
| $s$ | 1.1 |
| $CF(t)$ | $4.35 \cdot 10^7$ |
| $SumDCF(t)$ | $2.21 \cdot 10^7$ |
| $S_L(t,k)$ | 0.711 |
| $A_{SS}(t)$ | 0.895 |
| $T_{PH}$ | 280.38 |

**Table 6.7:** *Optimal Values of the Decision Variables to maximize $CF(t)$*

By comparing the optimized solutions for maximizing $CF(t)$ and $SumDCF(t)$ (see table 6.9), the time value of money causes a drift towards earlier preventive maintenance and repair activities on a higher quality level. The maintenance strategy maximizing $SumDCF(t)$ is represented in Figure 6.11.

## 6.6. Optimization

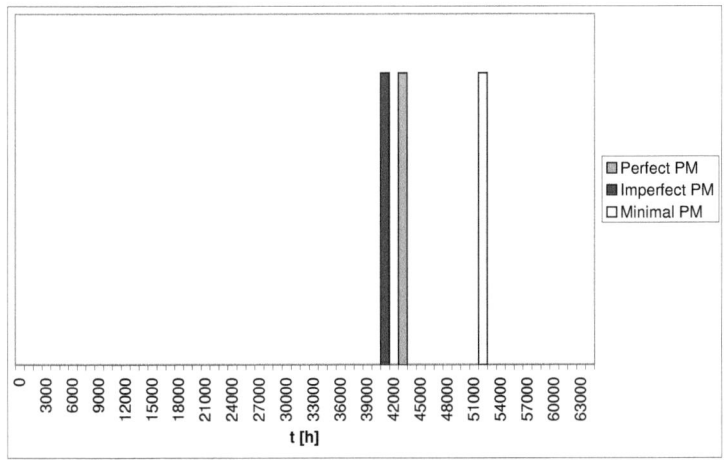

**Figure 6.10:** *Optimal Maintenance Strategy for Maximizing $CF(t)$*

| | |
|---|---|
| $I_{PM}$ | 64'000 |
| $I_{IM}$ | 52'700 |
| $I_{MM}$ | 49'800 |
| $P_{ImperfectR}$ | 1.00 |
| $P_{MinimalR}$ | 0.01 |
| $s$ | 1.1 |
| $CF(t)$ | $4.34 \cdot 10^7$ |
| $SumDCF(t)$ | $2.22 \cdot 10^7$ |
| $S_L(t,k)$ | 0.734 |
| $A_{SS}(t)$ | 0.907 |
| $T_{PH}$ | 280.38 |

**Table 6.8:** *Optimal Values of the Decision Variables to maximize $SumDCF(t)$*

### 6.6.7 Comparison of Optimized Systems with the Default System

Table 6.9 shows the optimization results in a compact form to compare them with the default system and the default system without preventive maintenance.

Results indicate the tendency of over-maintenance of the default system (availabilities

# 140 Chapter 6. Simulation and System Optimization

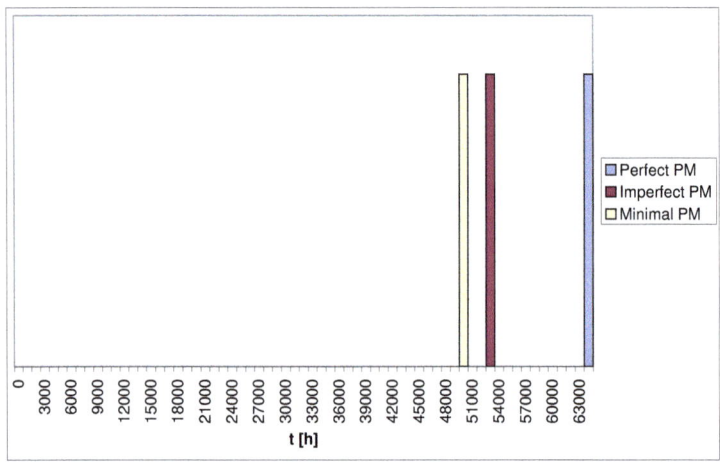

**Figure 6.11:** *Optimal Maintenance Strategy for Maximizing $SumDCF(t)$*

|  | Exp. 1 | Exp. 2 | Opt. 1 | Opt. 2 | Opt. 3 |
|---|---|---|---|---|---|
| $I_{PM}$ | – | - | 64'000 | 43'000 | 64'000 |
| $I_{IM}$ | 100 | - | 63'400 | 41'300 | 52'700 |
| $I_{MM}$ | – | - | 58'200 | 52'200 | 49'800 |
| $P_{ImperfectR}$ | 1.00 | 1.00 | 1.00 | 1.00 | 1.00 |
| $P_{MinimalR}$ | 0.00 | 0.00 | 0.41 | 0.11 | 0.01 |
| $s$ | 1.0 | 1.0 | 3.5 | 1.1 | 1.1 |
| $CF(t)$ | $4.20 \cdot 10^7$ | $4.18 \cdot 10^7$ | $2.16 \cdot 10^7$ | $4.35 \cdot 10^7$ | $4.34 \cdot 10^7$ |
| $SumDCF(t)$ | $2.10 \cdot 10^7$ | $2.09 \cdot 10^7$ | $1.10 \cdot 10^7$ | $2.21 \cdot 10^7$ | $2.22 \cdot 10^7$ |
| $S_L(t,k)$ | 0.005 | 0.006 | 0.978 | 0.711 | 0.734 |
| $A_{SS}(t)$ | 0.882 | 0.909 | 0.913 | 0.895 | 0.907 |
| $T_{PH}$ | 280.38 | 280.38 | 280.38 | 280.38 | 280.38 |

**Table 6.9:** *Comparison of optimized Systems with default System*

of all other system configurations are slightly better than the one of the default system). All optimizations expose a substantial lower preventive maintenance intensity than the original system. Furthermore, perfect repair activities denote an inconsiderable influence on $CF(t)$, $SumDCF(t)$ and $A_{SS}(t)$ since $P_{ImperfectR}$ is set to 1 and no perfect

## 6.7. Recommendation

repair activities are performed, thus.

Default system has some potential for optimization. The financial optimization gap is between an improvement of 3.5% in the case of maximizing the discounted cash flow and 5.7% for maximizing the cash flow. If the maximum system availability is in focus, $A_{SS}(t)$ can be increased by 3.5%. This potential is depicted in Figure 6.12 where all results of the default system are normalized to 100% and set in contrast to the optimized solutions.

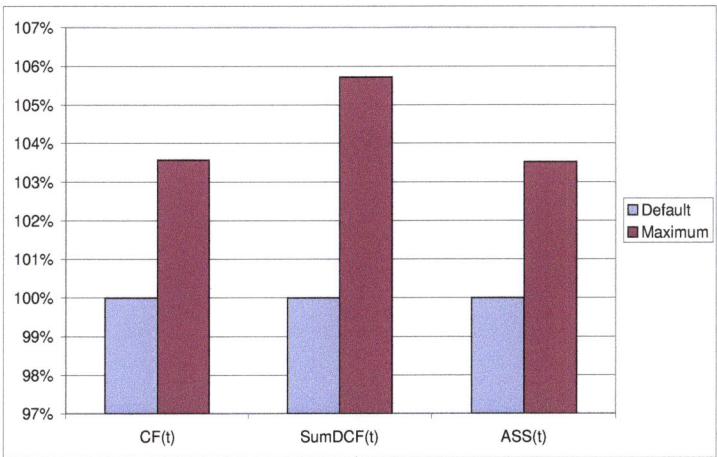

**Figure 6.12:** *Comparison between Normalized Results of the Default System and the Results of the Optimization*

### 6.7 RECOMMENDATION

Simulation and optimization experiments confirmed that the optimal maintenance strategy strongly depends on the chosen optimization criterion. Thus, the results greatly vary in their characteristics of sequential maintenance tasks and maintenance intensity. In general, the default system tends to be over-maintained, and it is recommended to reduce the amount of maintenance tasks. They reduce the system availability more than maintenance diminishes its proneness to fail. Moreover, when economic aspects come to the fore, the optimized system exposes a shift towards later preventive maintenance activities.

With respect to the quality level associated maintenance tasks defined in subsection 5.1.1.3, the optimized maintenance strategies are:

**Maximizing** $A_{SS}(t)$ Whenever system availability is in the focus of the optimization, a minimal maintenance activity should be executed at $t = 58'200$. Moreover, it is beneficial to perform only one imperfect preventive maintenance task at $t = 63'400$. The perfect preventive maintenance activity at the end of the optimization time should not be performed since it doesn't impact system availability but causes costs only.

   1. Replacement with spare parts of the same age at $t = 58'200$
   2. Do greasing etc, or eliminate weak-spots at $t = 63'400$

**Maximizing** $CF(t)$ Spotlighting the cash flow of the system leads to an increase in the maintenance activities. Since $CF(t)$ is an offspring of the measurand $SumDCF(t)$, it is not surprising that the associated optimized maintenance strategy is fairly similar. The difference between those two strategies is marginal. The optimized maintenance strategy for maximizing $CF(t)$ is graphically depicted in Figure 6.10

   1. Replacement with spare parts of the same age at $t = 52'200$
   2. Do greasing etc, or eliminate weak-spots at $t = 41'300$
   3. Do overhauling at $t = 43'000$

**Maximizing** $SumDCF(t)$ Integration of the aspect of time value of money causes a drift towards later maintenance activities than in the case of optimizing $CF(t)$. Associated optimal maintenance strategy is represented in Figure 6.11. The requested overhaul at the end of the lifetime should not be performed due to cost-reasons.

   1. Replacement with spare parts of the same age at $t = 49'800$
   2. Do greasing etc, or eliminate weak-spots at $t = 52'700$

Regarding the optimization of $CF(t)$ and $SumDCF(t)$, it could be stated that the associated optimal maintenance strategy depends on the investment appraisal approach used in the company to rate investments. Whenever static methods are applied, the

## 6.7. Recommendation

maintenance strategy that maximizes $CF(t)$ should be introduced. On the contrary, dynamic techniques promote the implementation of the maintenance strategy derived from optimizing $SumDCF(t)$.

The values of the financial objective functions of the corrective maintained system are all in the range of the optimized system. Thus, the efficiency of a further optimization going beyond corrective maintenance is in dispute. Indeed, choosing corrective maintenance and an increase of the safety factor would provide a sufficiently optimized system configuration, in this case. However, this conclusion could only be found by cross-checking all other alternatives (optimized systems).

Optimization runs have proven that financial goals do not coincide with the maintenance objective aimed at maximizing system availability. From the financial perspective, system availability's contribution to the financial success of a production system tends to be overrated. Furthermore, results show that sole availability considerations cannot be used as optimization criterion for maintenance strategy optimization since they provide an incomplete mapping of the situation. Availability and financial criteria as well as service level requirements should be simultaneously optimized to provide a well-balanced maintenance strategy.

However, in many situations the difficulty of multi-objective optimization can be overcome by replacing the objective functions of system availability and service level with a minimal boundary. Then, minimal system availability and service level can be integrated in the optimization procedure as constraints and the single optimization function is referred to financial aspects.

## Chapter 7
# Conclusions and Outlook

Advanced maintenance strategies and maintenance optimization have gained importance in the recent years. Mainly, two aspects have contributed to the raised awareness concerning maintenance:

- Complexity of modern production systems due to intensified data exchange between production systems and production planning and controlling applications.

- The spread of JiT and other concepts aimed at shortening lead times and minimizing stocks in combination with the customer's call for reliable product provision have aggravated the pressure on costs and production system availability.

Maintenance, as principle contributor to high system availability, is more and more considered to be a valuable supporter in meeting the customer's need of a high service level. The maintenance departments undergo a renunciation from a simple reactive cost center towards a service provider. Based on this problem area, the thesis is mainly aimed at:

1. Revealing the mutual impact of maintenance and logistics for proving for evidence the benefit of a joint optimization of logistics and maintenance.

2. Establishing an encompassing and expandable maintenance model providing quantitative measurands and incorporating availability, logistics, and financial aspects.

3. Optimizing the proposed model.

These goals have been achieved by:

1. Investigating analytically the interplay between system availability and the safety factor $s$ in logistics.

2. Pinpointing the interrelation between demand $P^{D(t,T_i)}$ and mission availability distribution $P^{A_{Mission_S}(t,k,cLT_j)}$ under the light of planning horizon $T_{PH}$ determination. It is shown that the variance of the mission availability distribution is impacted by preventive maintenance.

3. Expanding the hybrid preventive maintenance model by integrating quality levels and applying failure rates subdivided into a maintainable and non-maintainable part.

4. Integration and modelling of failed maintenance under consideration of training and learning effects.

5. Using discounted cash flow as objective function to optimize maintenance strategies which provides an dynamic integration of the time value of money.

Simulation runs showed that the default system tends to be over-maintained and proved for evidence the dependency between safety factor and system availability.
In addition to the classic optimization criteria ,as minimizing costs and maximizing system availability, the overall cash flow and the discounted cash flow of the production system were taken as supplementary objective functions. In particular, the discounted cash flow offers an optimization over the whole lifetime of a production system.

Results of the optimization runs confirmed the assumption that system availability alone is an insufficient objective function for optimizing a maintenance strategy. Availability considerations have to be merged with financial aspects to achieve an optimal maintenance strategy that satisfies both, the customer and the production system operator.

Furthermore, these trends promote and highlight the meaningfulness of maintenance in optimizing the operation of production systems and call for sophisticated accounting methods to measure not only the costs, but also the benefit of maintenance activities. In particular, the estimation of the cost-effectiveness of maintenance tasks and maintenance strategies is still a challenge and an unresolved problem. This also refers to organisational aspects concerned with maintenance. Balanced scorecards (BSC) are an attempt to monitor and control the effectiveness of different maintenance strategies by quantifying qualitative assets in maintenance.

Although BSC has proven its strengths in combining quantitative and qualitative measurands in strategic planning and controlling, a large spread of this concept in maintenance is lacking yet. One of the reasons may be in the fact that this semi-quantitative approach is still too imprecise for sound and reliable cost-benefit considerations. As long as no proper quantification for this problem can be found, cost-benefit considerations and optimization concerning financial aspects in maintenance cannot cover issues beyond a semi-quantitative level. However, deficiencies in assessing the cost-efficiency of a maintenance strategy are highlighting only the financial aspects of maintenance strategy optimization.

Traditional maintenance optimization focuses on improving production system availability too.

The drift towards JiT concepts in logistics, where deliverability is impaired by production system availability, promotes the tendency to optimize rather the service level than the production system availability. For this reason, the interconnections between system availability and its impact on logistic objectives were investigated by stochastic modelling. Interdependencies between safety factor $s$ and system mission availability $A_{Mission_S}(t, k, cLT_j)$ on the service level $S_L(t, k)$ and fill rate $F_R(t, k)$ are demonstrated. Resulting equations could be used to estimate a lower boundary for the safety factor which is controlled by the production system utilization $U(t, cLT_j)$. Moreover, an interplay between system availability $A_{SS}(t, T_i)$ and the safety factor $s$ could derived. System availability is impacted by the chosen safety factor. This is surprising since it is obscure that a logistic control parameter may have an influence on a technical measurand.

Another interplay between logistics and system availability could be revealed; logistic equation connects demand with production output and inventory. Due to the fact that demand needs to be forecasted, this requires a forecasting of the production system output as well. Thus, all conditions regarding demand forecasting have to be applied for production system output prediction, too. Since deviation of production system output is influenced by the variability of the mission availability and the utilization of the production system in principle, production output planning reduces to manage mission availability and utilization. Utilization can be used as a control parameter and mission availability of the production system is a random variable. Thus, the same conditions to be met for forecasting are in force for both, demand and mission availability. Those conditions mainly concern a minimal planning horizon which should guarantee that both random variables, demand and mission availability of the production system, follow a normal distribution (coefficient of variance of the associated distributions should be below 0.4). Examples have shown that this coefficient of variance of the mission availability distribution is strongly affected by preventive maintenance activities. An adequately preventively maintained system exposes a lower coefficient of variance of its mission availability distribution than the same system with corrective maintenance only. This is relation between planning horizon and mission availability hasn't been demonstrated before and is a strong indicator that maintenance and logistics should be optimized jointly.

Further work was concerned with reviewing existing preventive maintenance models and different attempts to model and derive the failure rate shape of a production system. It has been proven for evidence, that preventive maintenance is only beneficial in the case of an increasing failure rate. Since most production systems are successfully maintained by applying the concept of preventive maintenance, it could be reasoned that they follow an increasing failure rate. Based on these findings, an integrative model incorporating the aspects of maintenance, impact of maintenance on the failure rate and logistics was elaborated.

The core element of the created model is the maintenance section in which the concept of the hybrid preventive maintenance model in combination with different quality levels of the associated maintenance tasks is realized. Whereas the classic hybrid preventive maintenance model is limited to minimal, imperfect and perfect preventive maintenance, the proposed model is extended to represent failed maintenance activities. For this reason, a maintenance induced failure model is introduced with a decreasing failure rate that covers the training effect. Another novelty in the maintenance model are the quality level related sojourn times in the associated states. Duration of repair and maintenance activities are modelled according to the quality level. Thus, it is assumed that a maintenance activity of higher quality requires more time than the execution of a maintenance task on a lower quality level. Since the proposed maintenance model is not only addressed to be optimized towards availability objectives but accounts for financial assets also, a maintenance cost rate is introduced to calculate the related maintenance costs. The maintenance module is embedded between a production and a logistics model which are concerned with the literal production process and the order dispatching. Although the production process is regarded as a black box, in particular, some specialities have been integrated. At the beginning of every new order, the production system needs a setup first. This setup is represented as an auxiliary delay in starting the production. Moreover, labor costs and costs for operating resources are introduced to account for the production costs.

Order dispatching and evaluation of the objective measurand is performed in the logistic model. Many models have tried to optimize maintenance strategies under consideration of demand, but none of these attempts have integrated the safety factor. This safety factor provides a higher degree of service level; a novelty in optimizing maintenance strategies. Maintenance strategies have rarely been optimized under the light of logistic objectives, as service level or fill rate. Implementation of the safety factor offers an integration of order rejecting. An order, whose planned finishing time is below the maximum planned finishing time of an order in the queue, is rejected, to provide a high service level. This unsatisfied demand is fined with a penalty. However, not only unsatisfied demand is imposed with a fine, but also delayed delivery. Those opportunity costs are integrated into the overall cost and cost-benefit calculations in the model.

It is an encompassing model that is able to represent some new aspects as the combination of logistics and maintenance which have never been integrated before. Furthermore, the subdivision of the model into a maintenance, production, and logistic part provides easy extendability and increase of degree of itemization.

Nevertheless, there are some possibilities for further work and expansions of the model.

**Maintenance Model** A vast variety of conceivable extensions of the model exists: First of all, the system failure rate could be modelled as an aggregation of several failure rates depending on the failure mechanisms of the sub-components of the production system to provide more accurate recommendations (which kind of maintenance activities on which part of the production system should be performed). In addition, the fixed time-steps between two consecutive maintenance tasks of the same quality level could be replaced by variable maintenance intervals to cope with the temporal hazard of incoming orders and failure occurrence. Moreover, the timely trigger of performing a maintenance activity can be changed into a performance driven indicator, meaning that after a certain amount of produced goods a maintenance activity should be executed. Such an additional flexibility in the setting of maintenance intervals may provide better results in comparison to fixed time-steps.

**Production Model** The literal production process is designed as a black box without a sufficiently higher degree of detailing. By segmenting the production process into its sub-processes and their associated components and workstations, the production process could be represented more realistically and would offer the possibility to integrate buffers and component/ workstation-specific failure rates or maintenance tasks. Furthermore, the representation of the actual layout of a production system allows to study the development of queues and to optimize buffer sizes or workstation cycle times to improve the output of the production system.

**Logistic Model** In the model, all of the demand is considered to be deterministic. However, in practice demand is mostly a combination of deterministic and stochastic demand whereas stochastic demand is based on demand forecasts. Thus, demand could be modelled as the superposition of deterministic and stochastic demand rather than assuming that all demand is deterministic. In this case, production and order dispatching could be better planned.

The simple order dispatching and rejecting mechanism can be replaced by a sophisticated dispatching methodology. There are many different planning techniques for order dispatching which can be divided into capacity- or order-oriented approaches. Although order-oriented planning procedures are of principle interest, since delivery in time is of essential importance in a JiT-logistics, a high utilization of the production system is required at the same time (financial considerations). Though, planning concepts as *Load-oriented Order Release (Loor)* or *Capacity-oriented materials management (Corma)* may neatly combine stochastic and deterministic demand to maximize utilization and service level.

**Optimization** The scatter search methodology built in the optimization engine $OptQuest^©$ could be expanded to a multidimensional optimization tool. Multidimensional search strategies, as Pareto Optimization, provide solutions for eventually concurring objective functions. This strategy could be applied for maximizing both, the service level $S_L(t,k)$ and the sum of discounted cash-flow $SumDCF(t)$. Furthermore, optimized solutions could be found under the condition of a minimum required service level $S_L(t,k)$.

# List of Tables

| | | |
|---|---|---|
| 3.1 | Minimum Safety Factor $s_{min}$ at Huba Control AG | 49 |
| 4.1 | Parameters for the Markov Chain with Increasing Failure Rate | 66 |
| 4.2 | Mean and Variance of Mission Availability with Increasing Failure Rate | 68 |
| 4.3 | Parameters for the Markov Chain with Constant Failure Rate | 69 |
| 4.4 | Mean and Variance of Mission Availability with Constant Failure Rate | 71 |
| 4.5 | Parameters for the Markov Chain with Decreasing Failure Rate | 72 |
| 4.6 | Mean and Variance of Mission Availability with Decreasing Failure Rate | 73 |
| 5.1 | Quality Levels of the PM Activities | 81 |
| 6.1 | Comparison of Simulation and Real Data | 123 |
| 6.2 | Simulation Results of $A_{SS}(t = 64'000)$ and $S_L(t = 64'000, k)$ | 123 |
| 6.3 | Simulation Results for $P^{A_{Mission_S}(k,p,T_i)}$ of Experiment 1 | 123 |
| 6.4 | Comparison between Preventive and Corrective Maintained Production Systems | 124 |
| 6.5 | Impact of $s$ on $CF(t)$, $SumDCF(t)$, $S_L(t,k)$ and $A_{SS}(t)$ | 126 |
| 6.6 | Optimal Values of the Decision Variables to maximize $A_{SS}(t)$ | 135 |
| 6.7 | Optimal Values of the Decision Variables to maximize $CF(t)$ | 136 |
| 6.8 | Optimal Values of the Decision Variables to maximize $SumDCF(t)$ | 137 |
| 6.9 | Comparison of optimized Systems with default System | 138 |
| A.1 | Modes of Interlinkage | 189 |

# List of Figures

1.1  Area of Conflict in the Field of Maintenance and Logistics . . . . . . . . . 2
1.2  Cause and Effect Diagram of Maintenance . . . . . . . . . . . . . . . . . . 3
1.3  Direct and Indirect Maintenance Costs . . . . . . . . . . . . . . . . . . . . 8
1.4  Time Value of Money . . . . . . . . . . . . . . . . . . . . . . . . . . . . . 11
1.5  Classification of Maintenance . . . . . . . . . . . . . . . . . . . . . . . . 14
1.6  OSA-CBM Framework . . . . . . . . . . . . . . . . . . . . . . . . . . . . 16
1.7  Temporal Measurands . . . . . . . . . . . . . . . . . . . . . . . . . . . . 19
1.8  Relation between Process and State Chart following [Meyer et al., 2006] . 24
2.1  Modelling Methods . . . . . . . . . . . . . . . . . . . . . . . . . . . . . . 30
2.2  Alternating Renewal Process . . . . . . . . . . . . . . . . . . . . . . . . . 35
2.3  State Diagram of the one Component, Renewable System . . . . . . . . . 37
3.1  Stocking Level according to [Schoensleben, 2002] . . . . . . . . . . . . . 43
3.2  Integral Intervals . . . . . . . . . . . . . . . . . . . . . . . . . . . . . . . 47
4.1  Categories of Failure Mechanisms . . . . . . . . . . . . . . . . . . . . . . 56
4.2  Definition of $\lambda_0(t)$ . . . . . . . . . . . . . . . . . . . . . . . . . . . . . . . 57
4.3  Hybrid Model . . . . . . . . . . . . . . . . . . . . . . . . . . . . . . . . . 58
4.4  Comparison of PM Models . . . . . . . . . . . . . . . . . . . . . . . . . . 59
4.5  Failure mechanisms . . . . . . . . . . . . . . . . . . . . . . . . . . . . . . 60
4.6  Semi-Markov Chain of Corrective Maintenance . . . . . . . . . . . . . . . 64
4.7  Semi-Markov Chain of Preventive and Corrective Maintenance . . . . . . 65

4.8  Comparison between System Availability with and without Preventive Maintenance for Increasing Failure Rate . . . . . . . . . . . . . . . . . . . . 66
4.9  Comparison between $\lambda_{Failure}(t)$ with and without Preventive Maintenance and Increasing Failure Rate . . . . . . . . . . . . . . . . . . . . . . . . . . 67
4.10 Histogram of System Mission Availability with and without Preventive Maintenance and Increasing Failure Rate . . . . . . . . . . . . . . . . . . 67
4.11 Comparison between System Availability with and without Preventive Maintenance for Constant Failure Rate . . . . . . . . . . . . . . . . . . . . 69
4.12 Comparison between $\lambda_{Failure}(t)$ with and without Preventive Maintenance and Constant Failure Rate . . . . . . . . . . . . . . . . . . . . . . . . . . . 70
4.13 Histogram of System Mission Availability with and without Preventive Maintenance and Constant Failure Rate . . . . . . . . . . . . . . . . . . . 70
4.14 Comparison between System Availability with and without Preventive Maintenance for Decreasing Failure Rate . . . . . . . . . . . . . . . . . . . 72
4.15 Histogram of System Mission Availability with and without Preventive Maintenance and Decreasing Failure Rate . . . . . . . . . . . . . . . . . . 73
4.16 Comparison between $\lambda_{Failure}(t)$ with and without Preventive Maintenance and Decreasing Failure Rate . . . . . . . . . . . . . . . . . . . . . . . . . 74

5.1  Simulation Model with Target Values printed in green, Influence Parameters labelled in red and Constraints in blue . . . . . . . . . . . . . . . . . 77
5.2  Simplified Maintenance Model as a State Chart . . . . . . . . . . . . . . 78
5.3  Data Flow Diagram of the Maintenance Model . . . . . . . . . . . . . . . 79
5.4  Preventive Maintenance Model as a State Chart . . . . . . . . . . . . . 80
5.5  Variable Definitions for the Preventive Maintenance Model . . . . . . . . 81
5.6  Instantaneous Effect of Preventive Maintenance on Failure Rate . . . . 82
5.7  Progression of the Failure Rate and Linear Approximation . . . . . . . . 83
5.8  Repair Model . . . . . . . . . . . . . . . . . . . . . . . . . . . . . . . . . 86
5.9  Variable Definition for the Repair Model . . . . . . . . . . . . . . . . . . 87
5.10 Transformation of $ProbSampleRepair$ . . . . . . . . . . . . . . . . . . . 88
5.11 Probability of Failure Prevention in Relation to Frequency of Preventive Maintenance Tasks according to [Anderson, 2002] . . . . . . . . . . . . 90

# List of Figures

5.12 Qualitative Failure Rate Progression before and after the first Preventive Maintenance Activity . . . . . . . . . . . . . . . . . . . . . . . . . . . . 98
5.13 Variable Definition of the Failure Model . . . . . . . . . . . . . . . . . . 99
5.14 Maintenance Model consists of the Adapted Hybrid Preventive Maintenance Model (see figure 5.4), the Repair Model 5.8 and the Failure Model 5.12 . . . . . . . . . . . . . . . . . . . . . . . . . . . . . . . . . . . . . 100
5.15 Variable Definition for Cost Calculation in the Maintenance Model . . . . . 101
5.16 Logistic Model . . . . . . . . . . . . . . . . . . . . . . . . . . . . . . . 103
5.17 Variable Definition for the Logistic Model . . . . . . . . . . . . . . . . . 104
5.18 Curve Fitting of Demand Quantity . . . . . . . . . . . . . . . . . . . . . 108
5.19 Goodness of Curve Fitting . . . . . . . . . . . . . . . . . . . . . . . . . 109
5.20 Object "DispatchedProductionOrder" . . . . . . . . . . . . . . . . . . . 110
5.21 Data Flow Diagram of the Production Model . . . . . . . . . . . . . . . 111
5.22 State Chart Production Controlling . . . . . . . . . . . . . . . . . . . . 112
5.23 Variable Definition for the Production Model . . . . . . . . . . . . . . . 114
5.24 Variable Definition for the Main Section . . . . . . . . . . . . . . . . . 115

6.1 Simulation Start Conditions . . . . . . . . . . . . . . . . . . . . . . . . 120
6.2 Comparison of Monthly System Availability . . . . . . . . . . . . . . . . 121
6.3 Scattergram of Real and Simulated Availability Values . . . . . . . . . . 122
6.4 Progression of $A_{SS}(t)$ . . . . . . . . . . . . . . . . . . . . . . . . . . 124
6.5 Progression of $S_L(t)$ . . . . . . . . . . . . . . . . . . . . . . . . . . . 125
6.6 Impact of different Safety Factors on Service Level . . . . . . . . . . . . 127
6.7 Impact of different Safety Factors on $CF(t)$ and $SumDCF(t)$ . . . . . . . 128
6.8 Optimization Plan . . . . . . . . . . . . . . . . . . . . . . . . . . . . . 129
6.9 Optimal Maintenance Strategy for Maximizing $A_{SS}(t)$ . . . . . . . . . . . 136
6.10 Optimal Maintenance Strategy for Maximizing $CF(t)$ . . . . . . . . . . . 137
6.11 Optimal Maintenance Strategy for Maximizing $SumDCF(t)$ . . . . . . . . 138
6.12 Comparison between Normalized Results of the Default System and the Results of the Optimization . . . . . . . . . . . . . . . . . . . . . . . . 139

# List of Figures

A.1 Venn Diagram for three sets $X_1$, $X_2$, and $X_3$ .................. 174

A.2 Curve approaches exact Value with Increase of incorporated Terms ... 174

A.3 Serial System of three Workstations ...................... 180

A.4 Simplified Functional Structure Diagram of an Interlinked Production System ......................................... 182

A.5 Tree Assembling .................................. 185

A.6 Network Assembling ............................... 186

A.7 Block Assembling ................................. 187

A.8 Influence of a Buffer on the Up-down Function ............... 189

A.9 Functional Structure Diagram of the Example ................ 190

A.10 Cumulative System Availability ........................ 191

B.1 Queueing System ................................. 193

D.1 RCM-Process .................................... 205

D.2 Overall Equipment Effectiveness ....................... 208

# Abreviations

| | |
|---|---|
| BDD | Binary Decision Diagram |
| BSC | Balanced Scorecard |
| CBM | Condition-based Maintenance |
| CCF | Common Cause Failures |
| CF | Cash Flow |
| CFR | Constant Failure Rate |
| CM | Corrective Maintenance |
| CV[X] | Coefficient of Variance of Distribution X |
| DFR | Descending Failure Rate |
| DSS | Decision Support System |
| FSD | Functional Structure Diagram |
| FT | Fault Tree |
| ICT | Information and Communication Technology |
| IFR | Increasing Failure Rate |
| iid | Independent Identical Distributed Property |
| IRR | Internal Rate of Interest |
| JiT | Just-in-Time |
| MLE | Maximum Likelihood Estimator |
| MTBF | Mean Time Between Failures |
| MTTR | Mean Time To Repair |
| NPV | Net Present Value |
| OEE | Overall Equipment Effectiveness |
| p | Interest Rate |
| PM | Preventive Maintenance |
| PPC | Production Planning and Control |
| RCM | Reliability-centered Maintenance |
| TPM | Total Productive Maintenance |

# Bibliography

[Abdulnour et al., 1995] Abdulnour, G., Dudek, R. A., and Smith, M. L. (1995). Effect of Maintenance on the Just-in-Time Production System. *International Journal of Production Ressources*, 33:565–583.

[Acel and Hrdliczka, 2002] Acel, P. and Hrdliczka, V. (2002). Simulation in Produktion und Logistik.

[Adam, 1989] Adam, S. (1989). *Optimierung der Anlageninstandhaltung*. Erich Schmidt Verlag.

[Al-Radhi, 2002] Al-Radhi, M. (2002). *TPM: Erfolgreich Produzieren mit TPM*. Hanser Verlag, München.

[Albino et al., 1992] Albino, V., Carella, G., and Okogbaa, O. G. (1992). Maintenance Policies in Just-in-Time Manufacturing Lines. *International Journal of Production Ressources*, 30:369–382.

[Alcade, 2000] Alcade, A. (2000). *Erfolgspotential Instandhaltung*. Erich Schmidt Verlag.

[Almog, 1979] Almog, R. (1979). *A Study of the Application of the Lognormal Distribution to Corrective Maintenance Repair Time Data*. PhD thesis, Naval Postgraduate School, Monterey, CA, USA.

[Anderson, 2002] Anderson, D. (2002). Reducing the Cost of Preventive Maintenance. Technical report, ONIQUA.

[Andradóttir, 1998] Andradóttir, S. (1998). A Review of Simulation Optimization Techniques. In *Proceedings of the 1998 Winter Simulation Conference*.

[April et al., 2003] April, J., Glover, F., Kelly, J. P., and Laguna, M. (2003). Practical Introduction to Simulation Optimization. In *Proceedings of the 2003 Winter Simulation Conference*.

[Axsaeter, 2006] Axsaeter, S. (2006). *Inventory Control*. Springer Verlag, Berlin.

[Bandow, 2006] Bandow, G. (2006). Nachhaltige Instandhaltung - Ergebnisse einer BMBF-Untersuchung. In *Dortmund*.

[Barlow and Proschan, 1975] Barlow, R. E. and Proschan, F. (1975). *Statistical Theory of Reliability and Life Testing*. Hol, Rinehart and Winston, New York.

[Birolini, 2007] Birolini, A. (2007). *Reliability Engineering: Theory and Practice*. Springer-Verlag, Berlin.

[Bishop, 2006] Bishop, C. M. (2006). *Pattern Recognition and Machine Learning*. Springer Verlag, Berlin.

[Block and Savits, 2001] Block, H. W. and Savits, T. H. (2001). Understanding the Shape of the Hazard Rate: A Process Point of View. *Statistical Science*, 16:14–16.

[Bodie and Merton, 2003] Bodie, Z. and Merton, R. C. (2003). *Finance*. Prentice Hall, Upper Saddle River.

[Bolch et al., 1998] Bolch, G., Greimer, S., de Meer, H., and Trivedi, K. S. (1998). *Queueing Networks and Markov Chains*. John Wiley & Sons, New York.

[Bollig and Wegener, 1996] Bollig, B. and Wegener, I. (1996). Improving the Variable Ordering of OBDDs Is NP-Complete. *IEEE Transactions on Computers*, 45:993–1002.

[Canfield, 1986] Canfield, R. V. (1986). Cost Optimization of Periodic Preventive Maintenance. *IEEE Transactions in Reliability*, 35:78–81.

[Carmichael, 1987] Carmichael, D. G. (1987). Machine Interference with General Repair and Running Times. *Mathematical Methods of Operations Research*, 31:115–133.

[Chen, 2004] Chen, Q. (2004). *The Probability, Identification, and Prevention of Rare Events in Power Systems*. PhD thesis, Iowa State University, Ames, Iowa.

[Choi et al., 2006] Choi, J., Realff, M. J., and Lee, J. H. (2006). Approximate Dynamic Programming: Application to Process Supply Chain Management. In *AIChE Journal*.

[Cook, 1971] Cook, S. A. (1971). The Complexity of Theorem-Proving Procedures. In *Proceedings of the Third Annual ACM Symposium on Theory of Computing*.

# Bibliography

[Cormen et al., 2000] Cormen, T. H., Leiserson, C. E., Rivest, R. L., and Stein, C. (2000). *Introduction to Algorithms*. MIT Press, Cambridge, USA.

[Crocker, 1999] Crocker, J. (1999). Effectiveness of Maintenance. *Journal of Quality in Maintenance Engineering*, 5:307–314.

[Das et al., 2007] Das, K., Lashkari, R. S., and Sengupta, S. (2007). Machine Reliability and Preventive Maintenance Planning for Cellular Manufacturing Systems. *European Journal of Operational Research*, 183:162–180.

[Davidsson et al., 2005] Davidsson, P., Logan, B., and Takadama, K. (2005). *Multi-Agent and Multi-Agent-Based Simulation*. Springer Verlag, Berlin.

[de Ron and Rooda, 2005] de Ron, A. J. and Rooda, J. E. (2005). Equipment Effectiveness: OEE Revisited. *IEEE Transactions on Semiconductor Manufacturing*, 18:190–196.

[Deventer, 2006] Deventer, R. (2006). Alles im Griff-Verwendung von Neuronalen Netzen bei Condition Based Maintenance. *atp-Automatisierungstechnische Praxis, Oldenbourg Industrieverlag GmbH*, 1:12–15.

[Dey and Rao, 2005] Dey, D. and Rao, C. R. (2005). *Handbook of Statistics 25: Bayesian Thinking, Modeling and Computation*. Elsevier-Sciences.

[Dhillon, 2002] Dhillon, B. S. (2002). *Engineering Maintenance*. CRC Press, Boca Raton USA.

[Dhillon and Liu, 2006] Dhillon, B. S. and Liu, Y. (2006). Human error in maintenance: A review. *Journal of Quality in Maintenance Engineering*, 12:21–36.

[DIN-13306, 2001] DIN-13306 (2001). *Begriffe der Instandhaltung*. Beuth Verlag GmbH.

[DIN-31051, 2001] DIN-31051 (2001). *Grundlagen der Instandhaltung*. Beuth Verlag GmbH.

[DIN-40041, 1990] DIN-40041 (1990). *Zuverlässigkeit: Begriffe*. Beuth Verlag GmbH.

[DIN-66201, 1981] DIN-66201 (1981). *Prozeßrechensysteme; Begriffe*. VDI Verlag GmbH.

[DIN-EN61078, 1994] DIN-EN61078 (1994). *Techniken für die Analyse der Zuverlässigkeit: Verfahren mit dem Zuverlässigkeitsblockdiagramm*. VDI Verlag GmbH.

[Donatelli, 1994] Donatelli, S. (1994). *Superposed Generalized Stochastic Petri Nets: Definition and Efficient Solution.* Springer Verlag, New York.

[Eichler, 1990] Eichler, C. (1990). *Instandhaltungstechniken.* Verlag Technik GmbH, Berlin.

[El-Ferik and Ben-Daya, 2006] El-Ferik, S. and Ben-Daya, M. (2006). Age-based Hybrid Model for Imperfect Preventive Maintenance. *IIE Transactions*, 38:365–375.

[Ergam, 1982] Ergam, C. (1982). *A Study of the Application of the Lognormal Distribution and Gamma Distribution to Corrective Maintenance Repair Time Data.* PhD thesis, Naval Postgraduate School, Monterey, CA, USA.

[Ezey, 2000] Ezey, D.-E. M. (2000). *Human Learning: From Learning Curves to Learning Organizations.* Kluwer Academic Publishers, Boston.

[Fabricius, 2003] Fabricius, S. (2003). *Modeling and Simulation for Plant Performability Assessment with Application to Maintenance in the Process Industry.* PhD thesis, ETH Zurich.

[Ferschl, 1964] Ferschl, F. (1964). *Zufallsabhängige Wirtschaftsprozesse.* Physica Verlag.

[Finkelstein and Esaulova, 2001] Finkelstein, M. S. and Esaulova, V. (2001). Why the Mixture Failure Rate decreases. *Reliability Engineering and System Safety*, 71:173–177.

[Fu, 2002] Fu, M. (2002). Optimization for Simulation: Theory and Practice. *INFORUS, Journal on Computing*, 14:192–215.

[Gani et al., 2003] Gani, J., Heyde, C. C., and Kurtz, T. G. (2003). *Probability and its Applications.* Springer Verlag, New York.

[Garey and Johnson, 1979] Garey, M. R. and Johnson, D. S. (1979). *Computers and Intractability: A Guide to the Theory of NP-Completeness.* W.H. Freeman.

[Gasmi et al., 2003] Gasmi, S., Love, C. E., and Kahle, W. (2003). A General Repair, Proportional-Hazards, Framework to Model Complex Repairable Systems. *IEEE Transactions on Reliability*, 52:26–32.

[Geering, 2001] Geering, H. P. (2001). *Regelungstechnik.* Springer Verlag, Berlin.

# Bibliography

[Gerencsér, 1999] Gerencsér, L. (1999). Optimization Over Discrete Sets via SPSA. In *Proceedings of the IEEE Conference on Decision and Control*.

[Gertler, 1998] Gertler, J. J. (1998). *Fault Detection and Diagnosis in Engineering Systems*. Marcel Dekker.

[Gharbi et al., 2007] Gharbi, A., Kenné, J.-P., and Beit, M. (2007). Optimal Safety Stocks and Preventive Maintenance Periods in Unreliable Manufacturing Systems. *International Journal of Production Economics*, 107:422–434.

[Ghiani et al., 2004] Ghiani, G., Laporte, G., and Musmanno, R. (2004). *Introduction to Logistics Systems Planning and Control*. John Wiley & Sons, Ltd., West Sussex.

[Glover and Laguna, 2002] Glover, F. and Laguna, M. (2002). *Handbook of Applied Optimization*, chapter Tabu Search, pages 194–208. Oxford University Press.

[Goldratt, 1999] Goldratt, E. M. (1999). *Theory of Constraints*. North River Press, Great Barrington.

[Guerkan et al., 1994] Guerkan, G., Oezga, A. Y., and Robinson, S. M. (1994). Sample-Path Optimization in Simulation. In *Proceeding of the 1994 Winter Simulation Conference*.

[Hackett, 1983] Hackett, E. A. (1983). Application of a set of learning curve model to repetitive tasks. *Radio und Electronic Engineer*, 53:25–32.

[Halevi, 2001] Halevi, G. (2001). *Handbook of Production Mangement Methods*. Butterworth-Heinemann, Oxford.

[Hermanns, 2002] Hermanns, H. (2002). *Interactive Markov Chains: The Quest for Quantified Quality*. Springer Verlag, Berlin.

[Joines and Roberts, 1998] Joines, J. A. and Roberts, S. D. (1998). Fundamentals of Object-Oriented Simulation. In *Proceedings of the 1998 Winter Simulation Conference*.

[Kastner and Dankl, 1992] Kastner, H. and Dankl, A. (1992). *Die optimale Instandhaltungs-Software für Ihr Unternehmen*. TÜV Media GmbH.

[Katoen, 1999] Katoen, J.-P. (1999). *Formal Methods for Real-Time and Probabilistic Systems*. Springer Verlag, New York.

[Kay, 1993] Kay, S. M. (1993). *Fundamentals of Statistical Signal Processing: Estimation Theory*. Prentice Hall, New Jersey.

[Kececioglu, 1991] Kececioglu, D. (1991). *Reliability Engineering Handbook*. Prentice Hall, New Jersey.

[Kececioglu, 1992] Kececioglu, D. (1992). *Reliability and Life Testing Handbook*. Prentice Hall, New Jersey.

[Kelly, 1997] Kelly, A. (1997). *Maintenance Strategy*. Butterworth-Heinemann, Oxford.

[Kenné et al., 2006] Kenné, J.-P., Gharbi, A., and Beit, M. (2006). Age-dependent Production Planning and Maintenance Strategies in unreliable Manufacturing Systems with lost Sale. *European Journal of Operational Research*, 178:408–420.

[Kijima, 1989] Kijima, M. (1989). Some Results for Repairable Systems with General Repair. *Journal of Applied Probability*, 26:248–370.

[Kijima and Suzuki, 1988] Kijima, M. and Suzuki, H. (1988). Periodical Replacement Problem without assuming Minimal Repair. *European Journal of Operations Research*, 37:194–203.

[Kirwan, 1994] Kirwan, B. (1994). *A Guide to Practical Human Reliability Assessment*. CRC Press, Boca Raton, USA.

[Konold and Reger, 2003] Konold, P. and Reger, H. (2003). *Praxis der Montagetechnik*. Vieweg Praxiswissen.

[Korn and Wait, 1978] Korn, G. A. and Wait, J. V. (1978). *Digital Continuous-System Simulation*. Prentice-Hall, Englewood Cliffs, New York.

[Kwon and Lee, 2004] Kwon, O. and Lee, H. (2004). Calculation Methodology for Contributive Managerial Effect by OEE as a Result of TPM Activities. *Journal of Quality in Maintenance Engineering*, 10:263–272.

[Laguna, 1997] Laguna, M. (1997). Optimization of Complex Systems with OptQuest.

[Laguna, 2002] Laguna, M. (2002). Scatter Search. In *Handbook of Applied Optimization*.

[Laguna, 2008] Laguna, M. (2008). E-Mail.

[Lehmann and Casella, 1998] Lehmann, E. L. and Casella, G. (1998). *Theory of Point Estimation*. Springer Verlag, Berlin.

# Bibliography

[Lewis, 1997] Lewis, C. D. (1997). *Demand Forecasting and Inventory Control: A Computer Aided Learning Approach.* John Wiley & Sons, New York.

[Lewis, 1987] Lewis, E. E. (1987). *Introduction to Reliability Engineering.* John Wiley & Sons, New York.

[Lie and Chun, 1986] Lie, C. H. and Chun, Y. H. (1986). An Algorithm for Preventive Maintenance Policy. *IEEE Transactions in Reliability*, 35:71–75.

[Lim et al., 2005] Lim, J.-H., Kim, D.-K., and Park, D. H. (2005). Cost Evaluation for an Imperfect-Repair Mode with Random Repair Time. *International Journal of Systems Science*, 36:717–726.

[Lin et al., 2001] Lin, D., Zuo, M. J., and Yam, R. C. M. (2001). Sequential Imperfect Preventive Maintenance Models with Two Categories of Failure Modes. *Naval Research Logistics*, 48:172–183.

[Lotter and Wiendahl, 2006] Lotter, B. and Wiendahl, H.-P. (2006). *Montage in der Industriellen Produktion.* Springer Verlag, Berlin.

[Luxhoj et al., 1997] Luxhoj, J. T., Riis, J. O., and Thorsteinsson, U. (1997). Trends and Perspectives in Industrial Maintenance Management. *Journal of Manufacturing Systems*, 16:137–453.

[MacKay, 2003] MacKay, D. J. C. (2003). *Information Theory, Inference and Learning Algorithms.* Cambridge University Press.

[Malik, 1979] Malik, M. A. K. (1979). Reliable Preventive Maintenance Schedule. *AIIE Transactions*, 11:221–228.

[Marin et al., 2005] Marin, J. M., Mengersen, K., and Robert, C. P. (2005). *Handbook of Statistics 25*, chapter Bayesian Modelling and Inference on Mixtures of Distributions, pages 1–50. Elsevier-Sciences.

[Meinel and Theobald, 1998] Meinel, C. and Theobald, T. (1998). *Algorithms and Data Structures in VLSI Design.* Springer Verlag, Berlin.

[Mendelson, 1997] Mendelson, E. (1997). *Introduction to Mathematical Logic.* Chapman & Hall, London.

[Messen and Mohr, 1982] Messen, H. and Mohr, G. (1982). Beschreibung und Bewertung von Personenhandlungen. Technical report, Kernforschungsanlage Juelich GmbH.

[Meyer et al., 2006] Meyer, U. B., Creux, S. E., and Weber Marin, A. K. (2006). *Process Oriented Analysis: Design and Optimization of Industrial Production Systems*. CRC Press, Taylor & Francis, Boca Raton.

[Mi, 1991] Mi, J. (1991). Interval Estimation of Availability of a Series System. *IEEE Transactions on Reliability*, 40:541–546.

[Mock, ] Mock, R. Instrumente zur Verfügbarkeitsbewertung integrierter technischer Systeme.

[Montgomery and Myers, 2002] Montgomery, D. C. and Myers, R. H. (2002). *Response Surface Methodology: Process and Product Optimization Using Designed Experiments*. John Wiley & Sons.

[Moubray, 1991] Moubray, J. (1991). *Reliability-Centered Maintenance*. Butterworth-Heinemann, Oxford.

[Moubray, 1996] Moubray, J. (1996). *RCM-Die hohe Schule der Zuverlässigkeit von Produkten und Systemen*. Verlag Moderne Industrie, Landsberg.

[Mullen, 2006] Mullen, R. (2006). Characterizing Software Defect Repair Times. In *The 17th IEEE International Symposium on Software Reliability Engineering, ISSRE 2006*.

[Nagakawa, 1986] Nagakawa, T. (1986). Periodic and Sequential Preventive Maintenance Policies. *Journal of Applied Probability*, 23:536–542.

[Nagakawa, 1988] Nagakawa, T. (1988). Sequential Imperfect Preventive Maintenance Policies. *IEEE Transactions in Reliability*, 37:295–298.

[Nakajima, 1988] Nakajima, S. (1988). *Introduction to TPM*. Productivity Press, Portland.

[Navarro and Hernandez, 2004] Navarro, J. and Hernandez, P. J. (2004). How to obtain Bathtube-Shaped Failure Rate Models from Normal Mixtures. *Probability in Engineering and Informational Sciences*, 18:511–531.

[Nelson, 1997] Nelson, W. R. (1997). Integrated Desing Environment for Human Performance and Human Reliability Analysis. In *IEEE Sixth Annual Human Factors Meeting*.

[Neuhaus, 2007] Neuhaus, H. (2007). Instandhaltung schafft Werte; Der Beitrag der Instandhaltung zur Wertschöpfung. In *Forum Vision Instandhaltung e. V., ÖVIA-Kongress*.

[Nowlan and Heap, 1978] Nowlan, F. S. and Heap, H. F. (1978). Reliability-Centered Maintenance, Tech. Rep. AD/A066-579. Technical report, National Technical Informaiton Service, US Department of Commerce, Springfield, Virginia.

[Nusbaumer, 2007] Nusbaumer, O. P. M. (2007). *Analytical Solutions of Linked Fault Tree Probabilistic Risk Assessments using Binary Decision Diagrams with Emphasis on Nuclear Safety Applications*. PhD thesis, ETH Zurich.

[O'Connor, 1995] O'Connor, L. (1995). The Inclusion-Exclusion Principle and its Applications to Cryptography.

[Osaki, 2002] Osaki, S. (2002). *Stochastic Models in Reliability and Maintenance*. Springer Verlag, Berlin.

[Papula, 1999] Papula, L. (1999). *Mathematik für Ingenieure und Naturwissenschaftler, Band 3*. Vieweg, Braunschweig, Wiesbaden.

[Pekka, 2000] Pekka, P. (2000). An Analysis of Maintenance Failures at a Nuclear Power Plant. *Reliability Engineering and System Safety*, 71:293–302.

[Perkins and Srikant, 2001] Perkins, J. R. and Srikant, R. (2001). Failure-Prone Production Systems with Uncertain Demand. *IEEE Transactions on Automatic Control*, 46:441–449.

[Pham, 2003] Pham, H. (2003). *Handbook of Reliability Engineering*. Springer Verlag, Berlin.

[Pinjala et al., 2006] Pinjala, S. K., Pintelton, L., and Vereecke, A. (2006). An Empirical Investigation on the Relationship betweeen Business and Maintenance Strategies. *International Journal of Production Economics*, 104:214–229.

[Pocock et al., 2001] Pocock, S., Harrison, M., Wright, P., and Johnson, P. (2001). THEA: A Technique for Human Error Assessment early in Design. Technical report, University of York, Heslington, U.K.

[Podratz, 2007] Podratz, K. (2007). Trend in der Instandhaltung. In *VTH TOP-Partner Forum, "Technischer Handel: Industrielle Instandhaltung-nur mit uns!"*.

[Rausand, 1998] Rausand, M. (1998). Reliability Centered Maintenance. *Reliability Engineering and System Safety*, 60:121–132.

[Reason, 1990] Reason, J. (1990). *Human Error*. Cambridge University Press.

[Reiman, 1994] Reiman, L. (1994). *Expert Judgment in Analysis of Human and Organziational Behavior at Nuclear Power Plants*. PhD thesis, Finnish Centre for Radiation and Nuclear Safety.

[Riah-Belkaoui and Holzer, 1986] Riah-Belkaoui, A. and Holzer, H. P. (1986). *The Learning Curve: A Management Accounting Tool*. Quorum Books, Westport, CT.

[Ritter and Schooler, 2002] Ritter, F. E. and Schooler, L. J. (2002). *International Encyclopedia of the Social and Behavioral Sciences*, chapter The Learning Curve, pages 8602–8605. Pergamon, Amsterdam.

[Roberts et al., 1981] Roberts, N., Vesely, W., Haasl, D., and Goldberg, F. (1981). *Fault Tree Handbook (NUREG-0492)*. U.S. Nuclear Regulatory Commission.

[Samanta et al., 1985] Samanta, P. K., O'Brian, J. M., and Morrison, H. W. (1985). *Multiple - Sequential FAilure Model: Evaluation and Procedures for Human Failure Dependency*. NUREG/CR-3637, Brookhaven National Laboratory.

[Sánchez et al., 2006] Sánchez, A. I., Martínez-Alzamora, N., and Mullor, R. (2006). Parameters Estimation under Preventive Imperfect Maintenance. In *ESREL06*.

[Savsar, 1997] Savsar, M. (1997). Simulation Analysis of Maintenance Policies in Just-In-Time Production Systems. In *International Journal of Operations & Production Management*.

[Schoensleben, 2001] Schoensleben, P. (2001). *Integrales Informationsmanagement*. Springer Verlag, Berlin.

[Schoensleben, 2002] Schoensleben, P. (2002). *Integrales Logistikmanagement*. Springer-Verlag, Berlin.

[Seiler, 2000] Seiler, A. (2000). *Financial Management, BWL in der Praxis II*. Orell Füssli Verlag AG.

[Sheu et al., 2006] Sheu, S.-H., Lin, Y.-B., and Liao, G.-L. (2006). Optimum Policies for a System with General Imperfect Maintenance. *Reliability Engineering and System Safety*, 91:362–369.

# Bibliography

[Shirmohammadi et al., 2007] Shirmohammadi, A. H., Zhang, Z. G., and Love, E. (2007). A Computational Model for Determining the Optimal Preventive Maintenance Policy with Random Breakdowns and Imperfect Repairs. *IEEE Transactions on Reliability*, 56:322–339.

[Siegel et al., 1984] Siegel, A. I., Bartter, W. D., Wolf, J. J., Knee, H. E., Haas, H. E., and Haas, P. M. (1984). *Maintenance Personnel Performance Simulation (MAPPS) Model*. NUREG/CR-3626.

[Siegle, 1995] Siegle, M. (1995). *Beschreibung und Analyse von Markovmodellen mit grossem Zustandsraum*. PhD thesis, Universität Erlangen.

[Smith, 1993] Smith, A. M. (1993). *Reliability-Centered Maintenance*. McGraw-Hill, New York.

[Smith and Tate, 1998a] Smith, H. and Tate, A. (1998a). Maintenance Programme Set-Up. Technical report, BSRIA.

[Smith and Tate, 1998b] Smith, H. and Tate, A. (1998b). Maintenance Programme Set-Up. Technical report, BSRIA.

[Soyer et al., 2004] Soyer, R., Mazzuchi, T. A., and Singapurwalla, N. D. (2004). *Mathematical Reliability: An Expository Perspective*. Springer Verlag, Berlin.

[Spall, 2003] Spall, J. C. (2003). *Introduction to Stochastic Search and Optimization Estimation, Simulation, and Control*. John Wiley & Sons, New York.

[Stephens, 2000] Stephens, M. (2000). Dealing with Label Switching in Mixture Modes. *Journal of the Royal Statistical Society. Series B (Statistical Methodology)*, 62:795–809.

[Suzuki, 1994] Suzuki, T. (1994). *TPM in Process Industry*. Productivity Press, Portland.

[Swain and Guttmann, 1983] Swain, A. D. and Guttmann, H. E. (1983). *Handbook fo Human Reliability Analysis with Emphasis on Nuclear Power Plant Applications*. NUREG/CR-1278, Sandia National Laboratories, Albuquerque, USA.

[Taylor, 2004] Taylor, J., W. (2004). Smoot Transition Exponential Smoothing. *Journal of Forecasting*, 23:385–394.

[Towill, 1990] Towill, D. R. (1990). Forecasting learning curves. *International Journal of Forecasting*, 6:25–38.

[Trigg and Leach, 1967] Trigg, D. W. and Leach, A. G. (1967). Exponential smoothing with an adaptive response rate. *Operations Research*, 18:53–57.

[Vaurio, 2001] Vaurio, J. K. (2001). Modelling and Quantification of Dependent Repeatable Human Errors in System Analysis and Risk Assessment. *Reliability Engineering and System Safety*, 71:179–188.

[VDI-3633-Blatt-1, 1993] VDI-3633-Blatt-1 (1993). Simulation von Logistik-, Materialfluss- und Produktionssystemen (Grundlagen). VDI Verlag GmbH.

[VDI-3649, 1992] VDI-3649 (1992). Anwendung der Verfügbarkeitsrechnung für Förder- und Lagersysteme. Beuth Verlag GmbH.

[VDI-4008-Blatt-2, 1998] VDI-4008-Blatt-2 (1998). Boolesches Model. Beuth Verlag GmbH.

[VDI-4008-Blatt-7, 1986] VDI-4008-Blatt-7 (1986). Strukturfunktion und ihre Anwendung. VDI-Verlag GmbH.

[VDI-4008-Blatt-8, 1984] VDI-4008-Blatt-8 (1984). Erneuerungsprozesse. Beuth Verlag GmbH.

[Vogel, 2003] Vogel, S. (2003). *Teil II, Mathematische Statisitk, In: Stochastik für Informatiker*. TU Ilmenau.

[Wang and Lee, 2001] Wang, F. K. and Lee, W. (2001). Learning Curve Analysis in Total Productive Maintenance. *Omega*, 29:491–499.

[Warnecke, 1992] Warnecke, H. J. (1992). *Instandhaltung*. Verlag TÜV Rheinland GmbH.

[Warnecke et al., 1996] Warnecke, H.-J., Bullinger, H.-J., and Hichert, R. (1996). *Wirtschaftlichkeitsrechnung für Ingenieure*. Munich.

[Watson and Head, 2007] Watson, D. and Head, A. (2007). *Corporate Finance: Principles & Practice*. Pearson Education.

[Wellner, 2003] Wellner, K. (2003). *Entwicklung eines Immobilien-Portfolio-Management-Systems: Zur Optimierung von Rendite-Risiko-Profilen diversifizierter Immobilien-Portfolios*. BoD-Books on Demand.

[Willmott and McCarthy, 2001] Willmott, P. and McCarthy, D. (2001). *TPM - A Route to World-Class Performance*. Butterworth-Heinemann, Oxford.

# Bibliography

[Winsberg, 1999] Winsberg, E. (1999). Sanctioning Models: The Epistemology of Simulation. *Special Issue of Science in Context*, 12:275–292.

[Wondmagegnehu, 2004] Wondmagegnehu, E. T. (2004). On the Behavior and Shape of Mixture Failure Rates from a Familiy of IFR Weibull Distributions. *Naval Research Logistics*, 51:491–500.

[Wunderlich, 2005] Wunderlich, W. O. (2005). *Hydraulic Structures: Probabilistic Approaches to Maintenance*. ASCE Publications, Reston, USA.

[Yang and Nachlas, 2001] Yang, S.-C. and Nachlas, J. A. (2001). Bivariate Reliability and Availability Modeling. *IEEE Transactions on Reliability*, 50:26–35.

[Zequeira and Bérenguer, 2005] Zequeira, R. I. and Bérenguer, C. (2005). Periodic Imperfect Preventive Maintenance with Two Categories of Competing Failure Modes. *Reliability Engineering and System Safety*, 91:460–468.

## Appendix A

# Novel Approach of Availability Modelling

Production system availability has dramatically gained of importance due to higher work-load, shortened lead-times and reduced stock-keeping (widespread of JiT). Methods and procedures to calculate availability of production system are deviations of techniques coming from reliability, availability and safety engineering (compare with chapter 2). Availability and safety engineering is strongly linked to and influenced by businesses requiring high availability (off-shore businesses, process industries) or regulated sectors, as nuclear energy production or energy transportation. Due to absent adequate methods in manufacturing industries, where safety and availability are of deviated significance, these approaches were adapted without substantial changes.

One of their major drawbacks is their incapability to deal with simultaneous failures that lead to an explosion of combinational terms. State-of-the-art solution approaches try to manage this shortcoming by omitting coinstantaneous events. In the case of events with very low probability of occurrence, the likelihood of this combined event is negligible and the error is irrelevant if this term is ignored (rare event approximation). Since failure probabilities in interlinked production systems are significantly higher than in e.g. nuclear, their combined probability of simultaneous occurrence of failures has to be taken into account. Without this correction, system availability is underestimated.

Some approximation techniques exist to deal with simultaneous failures (Inclusion-Exclusion Principle [Smith and Tate, 1998b], Canonical Forms [Mendelson, 1997], Structure Function and Minimal Path Sets/Minimal Cut Sets [Birolini, 2007] but they are limited to static methods. Another deficiency is their shortcoming to represent temporal progression of system availability. System availability varies in time even when the failure probabilities of the underlying components are invariable. This time dependence is caused by the interplay of the different components (coinciding downtimes). Supply readiness and high service level [Schoensleben, 2002] has dramatically gained importance in manufacturing industries due to shortened lead-times claimed by the customers, Just-In-Time logistics [Savsar, 1997] and reduced or even no stock-keeping. In the case of no stockkeeping, high supply readiness and service level can only be guaranteed by high available production systems and reliable production planning (estimation of order dead-line). Production planning requires precise system availability calculations (essentially, good estimations about the variance of the production system availability) to avoid major errors in order deadline estimation.

## A.1 Approximation Techniques for dealing with Simultaneous Failures

One of the major drawbacks of the presented methods is their incapability to deal with simultaneous failures. Any methods using Boolean equations are facing the difficulty of over-exponential increase of combinational terms. In the Boolean equations, logic "AND" and "OR" need to be replaced with multiplications, summations and subtractions in order to be able to assign numerical values to the state variables.

In a disjunction of not mutually exclusive events, the set union of those events is not zero. These additional terms provoke a combinational term explosion. The difficulty of excessive increase of combinational terms belongs to the class of satisfiability problems. Satisfiability (SAT) is the issue of determining if the variables of a given Boolean formula can be assigned in such a way as to make the formula evaluate to true (see A.1.4). Discovered by Stephen Cook in 1971 [Cook, 1971], who proofed that SAT is NP-complete, this class of equation has been broadly discussed and still belongs to the most difficult problems in the field of NP-problems [Cormen et al., 2000]. NP-complete means that the problem is not solvable in polynomial time and is strictly limited to applications with a few variables, thus.

State-of-the-art solution approaches try to manage this shortcoming by omitting coinstantaneous events. This approximation is base on the fact that the probability of a combined event with two or more very unlikely events is negligible and the error is irrelevant if this term is ignored. Since failure probabilities in interlinked production systems are significantly higher than in e.g. nuclear energy production, their combined probability of simultaneous occurrence of failures has to be taken into account. Without this correction, system availability is underestimated (compare with equation A.2).

Progress in analysis methods and incorporating more and more (basis-) events have

## A.1. Approximation Techniques for dealing with Simultaneous Failures

lead to an increasing call for review of this formerly unattended combinations of rare events. Even analysts in the field of nuclear engineering are facing the fact that rare events have to be considered and introduced in their analysis. Tightened regulations and continuous improvement of the system reliability and availability have caused that formerly neglected basis events and event combinations have to be taken into considerations [Nusbaumer, 2007].
Some approximation techniques to deal with simultaneous failures will be discussed in the remainder of this chapter. Those methods can always be applied in situations referring to Boolean equations.

### A.1.1 Inclusion-Exclusion Principle

The Inclusion-exclusion (also known as the sieve principle) principle is one of the oldest methods used in probabilistic analysis that allows a probabilistic calculation of simultaneous events. The method belongs to a combinatorial problem of finding the number of objects having membership in a given subset [O'Connor, 1995].
The state variables $X_1, \ldots, X_n$ might be dependant and not disjoint. Than the probability of $P[X_1 \cup \ldots \cup X_n]$ is [Nusbaumer, 2007]:

$$P[X_1 \cup \ldots \cup X_n] = \sum_{1 \leq i \leq n} |X_i| - \sum_{1 \leq i_1 \leq i_2 \leq n} |X_{i_1} \cup X_{i_2}| + \ldots + (-1)^{n-1} |X_1 \cup \ldots \cup X_n| \quad (A.1)$$

The name comes from the idea that the principle is based on over-generous inclusion, followed by compensating exclusion. When $n > 2$ the exclusion of the pairwise intersections is (possibly) too severe, and the correct formula is as shown with alternating signs. Functionality of the theorem can graphically be demonstrated with Venn Diagrams (see Figure A.1). This method gives rise to approximate the exact value with infinitesimal error. Approximation approaches the exact value with increasing $i_1$ and $i_2$ (amount of incorporated subsets). The approximation until the $k^{th}$ combinational term is called $k^{th}$-order approximation.

Above equation A.1 can be modified to express system availability [Smith and Tate, 1998b] (see equation A.2).

$$A_{SS} = \prod_{i=1}^{n}(1 - P(X_i = 1))$$
$$+ \sum_{1 \leq i_1 \leq i_2 \leq n} P(X_{i_1} = 1 \cap X_{i_2} = 1) + \ldots + P(X_1 = 1 \cap \ldots \cap X_n = 1) \quad (A.2)$$

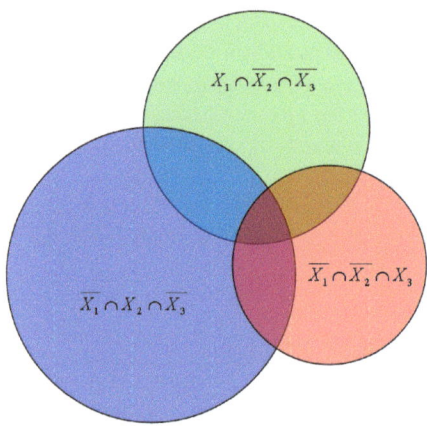

**Figure A.1:** *Venn Diagram for three sets* $X_1$, $X_2$, *and* $X_3$

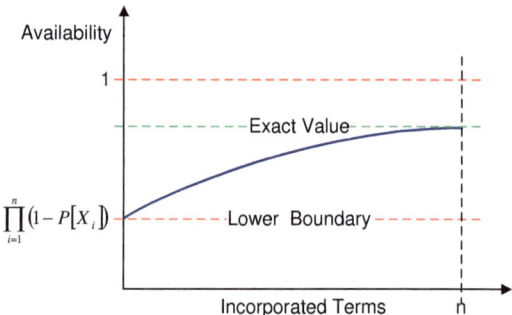

**Figure A.2:** *Curve approaches exact Value with Increase of incorporated Terms*

## A.1.2 Rare Event Approximation

Rare event approximation is a simplification of the inclusion-exclusion principle using only the first part of the equation A.2.

$$A_{SS} \;=\; \prod_{i=1}^{n}(1 - P(X_i = 1))$$

## A.1. Approximation Techniques for dealing with Simultaneous Failures

First-order rare event approximation neglects this residual part in disjunctions and sets

$$X \vee Y = P[X] + P[Y] \qquad (A.3)$$

to simplify calculation.

It makes use of the fact that if the individual probabilities of a term of events are infinitesimal, the calculation can be simplified by neglecting the higher order terms of the polynomial without loosing much accuracy [Chen, 2004].

### A.1.3 Structure Function Methods

It is assumed that the system is *coherent* [Birolini, 2007]. Coherence is understood as fulfilling the requirements:

1. Only active redundancy is taken into consideration.
2. Elements can appear several times in the model but are independent.
3. System state depends on the states of its elements.
4. If the system is in failure, no additional failure can reset it in up state (monotony).

Regarding requirements above, it can be stated that almost all systems are *coherent*. State space of a coherent system can be depicted with a system function $\Phi$ (structure function) and is a Boolean equation with the indicators $X_i = X_i(t)$. Every component $E_i$ has an associated state variable $X_i(t)$ with the properties:

$$X_i(t) = 0 \quad E_i \text{ failed}$$
$$X_i(t) = 1 \quad E_i \text{ not failed}$$

Then, the system function is defined as:

$$\Phi = \Phi(X_1, \ldots, X_n) = \begin{cases} 1 & \text{for system in up state} \\ 0 & \text{for system in down state} \end{cases}$$

for which the following holds:

1. $\Phi$ depends on $X_i$ with $(i = 1, \ldots, n)$
2. $\Phi = 0$ for all $S_i = 0$ and $\Phi = 1$ for all $X_i = 1$

This formulation of the structure function implicates that the availability function of the system can be expressed in terms of $\Phi$. System availability is:

$$A_{SS}(t) = P[\Phi(X_1(t),\ldots,X_n(t)) = 1] = E[\Phi(X_1(t),\ldots,X_n(t))]$$

Since calculation of $E[\Phi]$ is in general easier than computing $P[\Phi = 1]$, application of the structure function transfers the problem of calculating $P[\Phi = 1]$ to that of determining the system function $\Phi(X_1,\ldots,X_n)$.

The definition of the structure function gives rise to two simple approaches to get system availability function.

### A.1.3.1 Minimal Cut Sets

**Definition 1** *Minimal cut sets are the minimal set of failed entities $X_j = 0$ for all $E_j$ are element of $C_i$, which block the transition from the entry to the exit. [VDI-4008-Blatt-7, 1986].*

A cut set is a combination of events that can lead to a system failure. It's called to be minimal (irreducible) if, when any basic event is eliminated from the set, the remaining events collectively are no longer a cut set [Kececioglu, 1991]. $C_i$ is a minimal cut set if the system is down when $X_j = 0$ for all $E_j \in C_i$ and $X_k \ni C_i$, but this does not hold for any subset of $C_i$. Within $C_i$, the elements $E_j$ constitute a parallel model. A failure of all components in the minimal cut set triggers a system failure.

For a given system with $m$ minimal cut sets, the availability function is:

$$A_{SS}(t) = \Phi = \Phi(X_1,\ldots,X_n) = \prod_{i=1}^{m}\left(1 - \prod_{E_j \in C_i}(1 - X_j)\right)$$

Traditionally, minimal cut sets approach is used for complex Block Diagrams or Fault Trees to achieve estimations about failure probabilities or likelihood of top-event occurrence. Those complex structures cannot be simplified by a combination of simple constructs (parallel, series and k-out-of-n subsystems). Besides of providing estimations, minimal cut sets can be applied to understand the structural vulnerability of the system; the longer a minimal cut set is, the less vulnerable the system is to that event-chain. Also, the amount of minimal cut sets is correlated with the tendency towards vulnerability.

## A.1.3.2 Minimal Path Sets

**Definition 2** *Minimal path sets $\mathcal{P}_i$ are the minimum sets of working entities $X_j = 1$ for all $E_j \in \mathcal{P}_i$ and $X_k = 0$ for all $E_k \ni \mathcal{P}_i$, which keep the transition from entry to exit open. The system fulfills its function if there is at least on "open" path between the input and the output in which all entities are working [VDI-4008-Blatt-7, 1986].*

For a given system with $r$ minimal path sets, the corresponding availability function $A_{SS}(t)$ is:

$$A_{SS}(t) = 1 - \prod_{i=1}^{r} 1 - \Phi_{\mathcal{P}_i} = 1 - \prod_{i=1}^{r}\left(1 - \prod_{E_j \in \mathcal{P}_i} X_j\right) \quad (A.4)$$

Minimal cut sets and minimal path sets are often used for computing Fault Trees. Those sets are correlated in the way that the minimal cut set of a Fault Tree represents the complement to the minimal path set of the same logic tree.

## A.1.4 Approaches using Canonical Form

Evaluation of Boolean equations is extremely time-consuming and required computational power exponentially increases with soaring amount of components. Thus, a more economic representation of the Boolean equation would be desirable. Canonical form refers to the word "normal form" and is used in logic to describe statements in a standard way. In a normal form every variable appears only one time and the Boolean equation can be represented with either only "OR" or "AND" operators. Generally, calculation of Boolean equations which are in a normal form is far less time-consuming than evaluating the original function.

### A.1.4.1 Disjunctive Normal Form

A statement is in disjunctive normal form if it is a disjunction (sequence of "ORs") consisting of one or more disjuncts, each of which is a conjunction (minterm) of one or more literals (i.e., statement letters and negations of statement letters) [Mendelson, 1997]. A minterm $p$ [VDI-4008-Blatt-7, 1986] is a conjunction (Boolean multiplication) of $n$ variables:

$$p \;=\; X_1 \vee X_2 \ldots \vee X_n$$

$X_i$ is either $X_i$ or $\overline{X_i}$ (not negatived or negatived) for $i = 1, 2 \ldots, n$. A minterm has the characteristic that it is evaluated to 1 if one of its variables is 1. Every Boolean function can be expressed as a disjunction of minterms and is called *disjunctive normal form*:

$$f(X_1, X_2, \ldots, X_n) \;=\; \bigvee_{i=0}^{2^n - 1} c_i \wedge p_i$$

With

$\quad\quad p_i$ $\quad$ Minterms
$c_i = 1$ $\quad$ if $p_i$ belongs to the minterms in the disjunction
$c_i = 0$ $\quad$ if $p_i$ belongs to every other minterm

Function $f(X_1, X_2, \ldots, X_n)$ is evaluated to 1 if at least one minterm is not 0. All logical formulas can be converted into a disjunctive normal form. Whereas the evaluation of the disjunctive normal form is very easy, its formulation may be difficult. In some cases, conversion to disjunctive normal form can cause an exponential explosion of the formula.

### A.1.4.2 Conjunctive Normal Form

A formula is in conjunctive normal form if it is a conjunction (sequence of "ANDs") of terms, where a term is a disjunction (maxterm) of literals [Mendelson, 1997]. A maxterm $s$ [VDI-4008-Blatt-7, 1986] is a disjunction (Boolean summation) of $m$ variables:

$$s \;=\; X_1 \wedge X_2 \ldots \wedge X_m$$

$X_j$ is either $X_j$ or $\overline{X_j}$ (not negatived or negatived) for $j = 1, 2 \ldots, m$. Maxterm has the characteristic that it is evaluated to 0 if one of its variables is 0.

$$f(X_1, X_2, \ldots, X_n) \;=\; \bigvee_{j=0}^{2^m - 1} c_j \vee s_j$$

With

$s_j$ Maxterms
$c_j = 0$ if $p_j$ belongs to the minterms in the conjunction
$c_j = 1$ if $p_j$ belongs to every other maxterms

The whole function $f(X_1, X_2, \ldots, X_n)$ is evaluated to 0 if only one minterm $p_j$ is 0.

Every propositional formula can be converted into an equivalent formula that is in conjunctive normal form. This is the reason why proofs are often based on conjunctive normal forms. Translation of a Boolean equation into a disjunctive or conjunctive normal form can lead to an exponential explosion of the formula. This phenomenon of exponential growth belongs to the k-SAT problem [Garey and Johnson, 1979]. The k-SAT problem is the difficulty of finding a satisfying assignment to a Boolean equation expressed in conjunctive normal form such that each disjunction contains at most $k$ variables. For any $k > 2$ this problem is NP-complete. Although evaluation time of the equation can be minimized, it is still an inefficient approach due to the fact that original equation has to be translated into its normal form which is an NP-complete problem.

## A.2 NOVEL APPROACH

It combines the idea of *alternating renewal processes*, where system availability is represented as an n-times convolution of the Mean Time Between Failures ($MTBF_i$) and Mean Time To Repair ($MTTR_i$), with modelling methods used in *automatic control*. Iterating up- and downtimes form a step-function (see figure A.3) that switches between the values 0 (downtime) and 1 (uptime). This up- and downtime pattern is strictly deterministic and is time-independent (pattern does not change in time). The binary characteristic can be deployed to derive system response by a simple superposition of all component step-functions. Up- and downtime characteristic of a single component $i$ is imprinted in the *transfer function* $G_{Si}(t)$. The system response $Y_S(t)$ of a production system is the result of the underlying component failure characteristics, represented by transfer functions $G_{Si}(t)$, and the input signal $U_S(t)$ (see Figure A.4). This superposition of $Y_S(t)$ is a convolution with the result that simultaneous failures are inherently integrated.

$U_S(t)$ represents additional interruptions of the interlinked production system caused by external effects, as e.g. preventive maintenance, setup, or production holdup due to absent production orders.

# Appendix A. Novel Approach of Availability Modelling

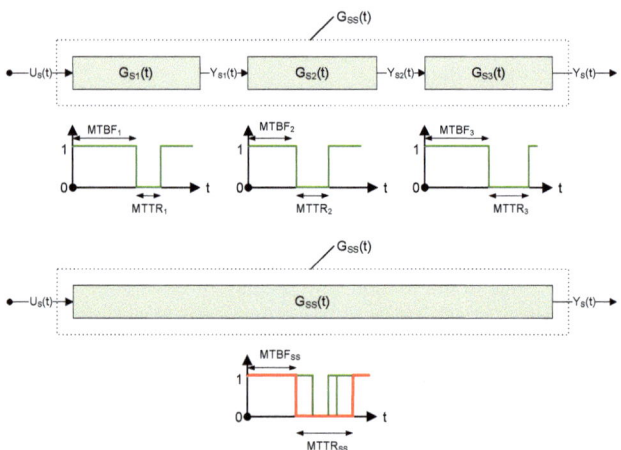

**Figure A.3:** *Serial System of three Workstations*

$$Y_S(t) = G_{SS}(t) * U_S(t) \qquad (A.5)$$

A simple computation of the system response $Y_S(t)$ requires a transformation of the stepwise-defined transfer function $G_{Si}(t)$ into a continuous and periodic function in time $t$. Means of choice is a Laplace transformation of the original function $G_{Si}(t)$ into the complex domain $s$ and a retransformation of the complex function $G_{Si}(s)$ into the time domain (see [Geering, 2001]).

## A.3 LAPLACE TRANSFORMATION

Laplace transformation offers an efficient method to transform a stepwise-defined function into a continuous, periodic one by transforming the given step-function from the time into a complex frequency domain.

The following equations base on the explanations in [Geering, 2001]. Fundamental Laplace transformation is defined in equation A.6.

$$\mathcal{L}\left[G_{Si}(t)\right] = G_{Si}(s) = -\frac{-1 + e^{-s \cdot MTBF_i}}{s \cdot \left(1 - e^{-s \cdot (MTBF_i + MTTR_i)}\right)}$$

## A.3. Laplace Transformation

Retransformed function $\mathcal{L}^{-1}[G_{Si}(s)]$ is:

$$\mathcal{L}^{-1}[G_{Si}(s)] = G_{Si}(t) = ceil\left(\frac{1}{MTBF_i + MTTR_i}t\right) \quad \text{(A.6)}$$
$$-floor\left(\frac{1}{MTBF_i + MTTR_i}t + \frac{MTTR_i}{MTBF_i + MTTR_i}\right)$$

$ceil$ = Smallest Integer greater than or equal to a number
$floor$ = Greatest Integer less than or equal to a number

### A.3.1 Serial System

A system has no redundancy if all components must work in order to fulfill the required function of the system. Equivalent network structure is a serial connection of all components of the system. It is assumed that each component operates and fails independently from every other component. Then, system transfer function $G_{SS}(t)$ of the serial system depicted in Figure A.3 is:

$$G_{Serial}(t) = G_{S1}(t) * G_{S2}(t) * \ldots * G_{Sl}(t) \quad \text{(A.7)}$$

In this particular case of a serial system where transfer functions take only binary values, the convolution of the system response $G_{SS}(t)$ is equal to a simple multiplication of the single component transfer functions. Cause for that exceptional effect is the binary characteristics of the component affecting system behavior in the way that an interruption of one component induces a system failure immediately (equation is already in a conjunctive normal form. Compare with subsection A.1.4.2). If the value 0 is assigned to one of the transfer functions $G_{Si}(t)$, the system transfer function $G_{SS}(t)$ immediately switches to 0. Thus, equation A.7 can be simplified.

$$G_{Serial}(t) = \prod_{i=1}^{l} G_{Si}(t) \quad \text{(A.8)}$$

Regarding the situation in Figure A.4, system is constituted by three subsystems in series. Internal structures of the other two subsystem types (parallel and k-out-of-n) can be replaced by and equivalent serial element that represents subsystem impact on system level. Thus, $G_{SS}(t)$ is the product of all subsystem transfer functions.

$$G_{SS}(t) = G_{Serial}(t) \cdot G_{Parallel}(t) \cdot G_{k \cap n}(t) \quad \text{(A.9)}$$

184    Appendix A. Novel Approach of Availability Modelling

Since all other subsystem types can be brought into a series system structure, convolution in the system response calculation (equation A.5) can also be substitute by a multiplication:

$$Y_S(t) = G_{SS}(t) \cdot U_S(t) \qquad (A.10)$$

Following [Birolini, 2007], system availability $A_{SS}(t)$ is :

$$A_{SS}(t) = \frac{1}{t} \cdot \int_0^t Y_S(x)dx \qquad (A.11)$$

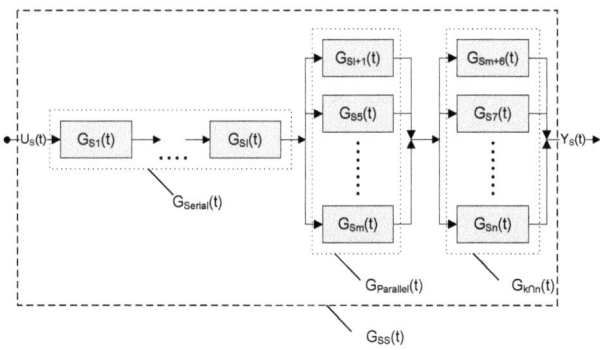

**Figure A.4:** *Simplified Functional Structure Diagram of an Interlinked Production System*

## A.4  Expansion to other Subsystem-Types

Figure A.4 is a deviation of a Block Diagram and offers a graphical depiction of the structural function in series, parallel, and k-out-of-n arrays. This notation shall be called Functional Structure Diagram (FSD) and uses block symbols and associated transfer functions to describe system failure characteristics. Even though a FSD resembles a structural layout of a system, it is only a representation of the inherent logical coherencies without any similarities with the physical assembly of the system.

## A.4.1 Parallel System

Whenever high available systems are required redundancy is a possibility to significantly increase and guarantee system availability. Redundancy is the existence of more than one single component for performing a function. (Active) redundancy of a component is a parallel system in a *FSD*. Operating mode of parallel systems is that the system fails when every redundant component fails. Formally, system is in operation when the sum of all transfer functions of the redundant components divided by the amount of redundancies $m - l + 1$ is in the range of $(0, 1]$.

$$G_{Parallel}(t) = ceil\left(\frac{1}{m-l+1} \cdot \sum_{i=l+1}^{m} G_{Si}(t)\right) \quad (A.12)$$

$m$ and $l$ are the indices of the parallel components (see Figure A.4).

## A.4.2 k-out-of-n System

A *k-out-of-n* subsystem can be depicted by encapsulated parallel subsystems and represents a specification of an ordinary parallel subsystem where $k$ components are necessary to keep the system operating and $n - k$ components are in active redundancy. $m$ and $n$ are the indices of the components in the *k-out-of-n* system (see figure A.4).

$$G_{k\cap n}(t) = ceil\left[\frac{1}{n} \cdot floor\left(\frac{1}{k} \sum_{i=m+1}^{n} G_{Si}(t)\right)\right] \quad (A.13)$$

Those three fundamental subsystem types are sufficient to adequately model most production systems. More sophisticated layouts are presented in the next section. *Although the problem area of maintenance crew availability and parallel maintenance tasks is not explicitly integrated in the proposed methodology, its impact on the system availability can be approximated by modifying $U_S(t)$.*

## A.5 ASSEMBLING LAYOUTS

Single components of a production system can be configured according to different linking and assembly principles. Well arranged material flow, high flexibility and a good

access to the components provide a sustainable cost effectiveness. Thus, system structure has a considerable impact on system productivity. After the system is set into operations, modifications on the system structure cause immense costs.
Regarding the principles of interlinkage in linear production systems, three methods can be distinguished [Lotter and Wiendahl, 2006]:

1. Decoupled Interlinkage
2. Elastic Interlinkage
3. Rigid Interlinkage

In a *rigid interlinkage*, workpieces are transported at the same time. Since there are no intermediate buffers between the components, an interruption of one component immediately causes a standstill of the whole system. To overcome this deficiency, buffers can be implemented between the components to compensate minor downtimes and losses due to idle time are reduced [Konold and Reger, 2003]. If buffer capacity is sufficient to completely decouple the components, this kind of interlinkage is called *decoupled interlinkage*. *Elastic interlinkage* is between those two examples and is designed to compensate only some minor interruptions. Enormous losses due to idle time in the case of rigid interlinkage and the costs for buffering in a decoupled interlinkage benefit the application of elastic interlinkage in practice. A short introduction into the field of buffers and queueing theory will be given in section A.6 and appendix B.

Layout and components assembling can be divided into:

- Line Assembling
- Tree Assembling
- Network Assembling
- Block Assembling

In *line assembling* all components are located along a transfer-line. Interlinkage between the components is either *rigid* or *elastic* and as long as no workpiece holder are used, which need to be returned to the start of the assembling line, the production line is an open system (a classic series system). If a return of the workpiece holder is required, the first and the last component are not completely decoupled anymore and approaches a block assembling (see subsection A.5.3).

## A.5. Assembling Layouts

*Tree* and *Network assembling* can only partially be deviated from the line assembling. Preliminary products of different component are locked in node points and are assembled in subsequent components in the case of a *tree structure*. Network structures are widespread in versatile productions where the outer components are used to assemble variant-specific parts which are final-assembled in the inner components.

*Block assembling* requires less space than the line assembling but access to the components is complicated and the mutual influence of the components increases.

Modelling of those assembling layouts with the proposed approach follows the idea of how sequential control systems are represented in automatic control.

### A.5.1 Tree Assembling

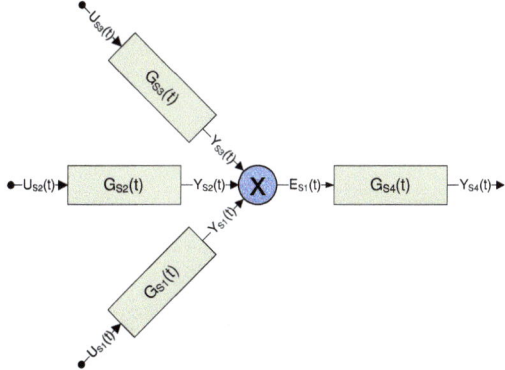

**Figure A.5:** *Tree Assembling*

$$\begin{aligned} Y_{S1}(t) &= U_{S1}(t) \cdot G_{S1}(t) \\ Y_{S2}(t) &= U_{S2}(t) \cdot G_{S2}(t) \\ Y_{S3}(t) &= U_{S3}(t) \cdot G_{S3}(t) \\ E_{S1}(t) &= Y_{S1}(t) \cdot Y_{S2}(t) \cdot Y_{S3}(t) \end{aligned}$$

$Y_{S4}(t)$ is:

$$Y_{S4}(t) = G_{S4}(t) \cdot \prod_{i=1}^{3} (U_{Si}(t) \cdot G_{Si}(t)) \qquad (A.14)$$

### A.5.2 Network Assembling

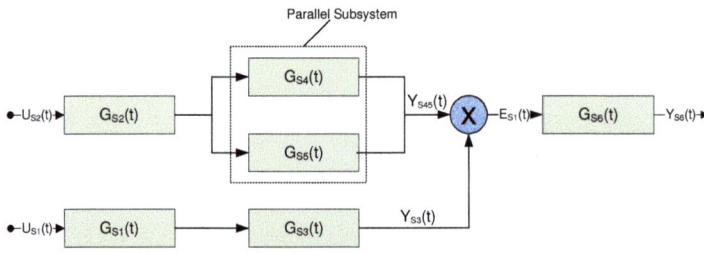

**Figure A.6:** *Network Assembling*

$$\begin{aligned} Y_{S45}(t) &= U_{S2}(t) \cdot G_{S2}(t) \cdot ceil\left(\frac{1}{2} \cdot (G_{S4}(t) + G_{S5}(t))\right) \\ Y_{S3}(t) &= U_{S1}(t) \cdot G_{S1}(t) \cdot G_{S3}(t) \\ E_{S1}(t) &= Y_{S3}(t) \cdot Y_{S45}(t) \end{aligned}$$

Than, $Y_{S6}(t)$ can be written as:

$$Y_{S6}(t) = G_{S6}(t) \cdot E_{S1}(t) \qquad (A.15)$$

### A.5.3 Block Assembling

$$\begin{aligned} E_{S1}(t) &= U_S(t) \cdot Y_S(t) \\ Y_S(t) &= E_{S1}(t) \cdot \prod_{i=1}^{8} G_{Si}(t) \end{aligned}$$

If the first equation is inserted in the second formula, resulting equation is:

$$Y_S(t) = U_S(t) \cdot Y_S(t) \cdot \prod_{i=1}^{8} G_{Si}(t)$$

## A.6. Buffers

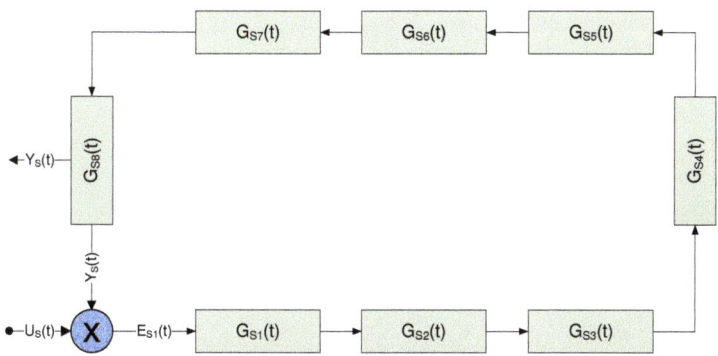

**Figure A.7:** *Block Assembling*

Equation A.16 can be divided by $Y_S(t)$. Then $Y_S(t)$ disappears in the formula.

$$U_S(t) \cdot \prod_{i=1}^{8} G_{Si}(t) = 1$$

The difficulty about this configuration is its interlinkage as a closed loop. Hence, this novel approach of availability modelling cannot be applied for block assembling. Other techniques as simulation or queueing networks can overcome the deficiencies of the proposed technique (see [Bolch et al., 1998]).
Those assembling layouts as all the other subsystems type are required to represent a production system into detail. They could be used to specify the black-box of the production model in section 5.3.

## A.6 BUFFERS

Buffer modelling and buffer characteristics are strongly linked with queueing theory. This theory is a discipline of operations research, introduced first by Agner Krarup Erlang in 1909 (see appendix B for a short introduction into queueing theory). A queue is a waiting line of jobs or work in a buffer waiting to be processed by a production system or component. Queues arise if the production rhythm of a system is not synchronized with its preceding and subsequent process steps (components). Buffers are installed between unsynchronized process steps to decouple processes and absorb potential disturbances in the production system. The result is a regular and continuous production flow.

There is no production type except continuous production in which all processes are completely synchronized. Thus, buffers exist in most production systems and are installed around critical or bottleneck capacities. This buffers provide that those capacities never run out of work and maximize the output of the system following the ideas of Theory of Constraints (see [Goldratt, 1999]) and its industrial application; the Drum-Buffer-Rope Concept.

### A.6.1 Buffer Modelling

Buffers affect component's transfer functions in the way that a buffer increases the ratio between up- and downtime by increasing $MTBF_i$, decreasing $MTTR_i$ but not affecting periodicity of the transfer function $G_{Si}(t)$. Buffering effect can be expressed as an added time interval $\triangle t_{Buffer_{i,i+1}}$ to $MTBF_i$ in which flow of supply to following components is not interrupted although one or several preceding components have failed (see figure A.8). During this bridging time $t_{Buffer_{i,i+1}}$ repair actions can already be performed. Therefore, the actual downtime due to repair is diminished and system availability is increased.

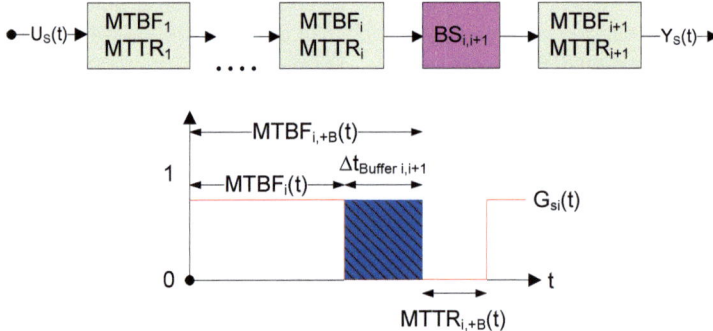

**Figure A.8:** *Influence of a Buffer on the Up-down Function*

Buffers may store outgoing products of a component if the subsequent component is blocked. Therefore, a buffer works always in both directions, up- and downstream.

## A.6. Buffers

$$\Delta t_{Buffer_{i,i+1}} = \frac{BS_{i,i+1}}{CP_{i+1}} \quad \text{(A.16)}$$

with

$BS_{i,i+1}$ = Buffer size of the buffer$_{i,i+1}$

$CP_i$ = Capacity of component $i$ $\left[\frac{pcs}{h}\right]$

Buffers have a bidirectional effect on work flow [VDI-3649, 1992]. A full buffer provides maximum safety against an interruption of the preceding component whereas an interruption of the subsequent component would cause a blocking of the preceding station and vice versa for an empty buffer. However, a blocked output-flow caused by a full buffer does not affect the output of the whole system due to already blocked buffer subsequent component. This is the reason why the downstream buffering effect is of interest only.

Equation A.16 offers a quantitative definition of the three principles of interlinkage; decoupled, elastic and rigid interlinkage.

The modified $MTBF_{i,+B}$ and $MTTR_{i,+B}$ are:

$$MTBF_{i,+B} = MTBF_i + \Delta t_{Buffer_{i,i+1}}$$
$$\approx MTBF_i + \frac{BS_{i,i+1}}{CP_{i+1}} \quad \text{(A.17)}$$
$$MTTR_{i,+B} = MTTR_i - \Delta t_{Buffer_{i,i+1}}$$
$$\approx MTTR_i - \frac{BS_{i,i+1}}{CP_{i+1}} \quad \text{(A.18)}$$

Those modified parameters are inserted in equation A.6.

$$G_{Si}(t) = ceil\left(\frac{1}{MTBF_{i,+B} + MTTR_{i,+B}} t\right) \quad \text{(A.19)}$$
$$-floor\left(\frac{1}{MTBF_{i,+B} + MTTR_{i,+B}} t + \frac{MTTR_{i,+B}}{MTBF_{i,+B} + MTTR_{i,+B}}\right)$$

| Decoupled Interlinkage | $t_{Buffer_{i,i+1}} \geq MTTR_i$ |
|---|---|
| Elastic Interlinkage | $0 < t_{Buffer_{i,i+1}} < MTTR_i$ |
| Rigid Interlinkage | $t_{Buffer_{i,i+1}} = 0$ |

**Table A.1:** *Modes of Interlinkage*

## A.7 Example

Let assume an interlinked production system, as described in Figure A.9, where $MTBF's$ and $MTTR's$ are given in minutes.

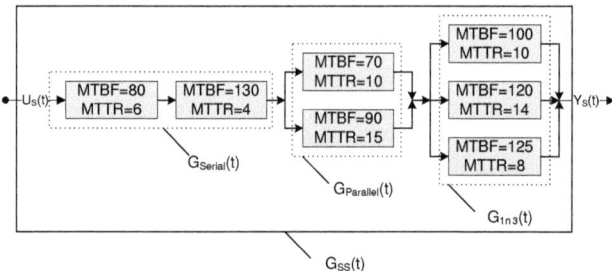

**Figure A.9:** *Functional Structure Diagram of the Example*

Equation A.6 can be applied to calculate the transfer functions of the components. As an example, this calculation is shown in detail for the first component on the left-hand side with $MTBF = 80[h]$ and $MTTR = 6[h]$:

$$G_{S1}(t) = ceil\left(\frac{1}{86}\right) - floor\left(\frac{t}{86} + \frac{6}{86}\right)$$

Subsystem transfer functions can be derived in an analog way by using equations A.10, A.12 and A.13. Equation A.20 depicts the result for the serial subsystem.

$$G_{Serial}(t) = \left(ceil\left(\frac{1}{86}\right) - floor\left(\frac{t}{86} + \frac{6}{86}\right)\right) \cdot \left(ceil\left(\frac{1}{134}\right) - floor\left(\frac{t}{134} + \frac{4}{134}\right)\right) \tag{A.20}$$

Transfer function $G_{SS}(t)$ of the whole system is:

$$G_{SS}(t) = G_{Serial}(t) \cdot G_{Parallel}(t) \cdot G_{1\cap 3}(t)$$

All external effects on the interlinked production systems are neglected ($U_S(t) = 1$) to simplify calculation. Then, equation A.10 provides system response:

$$Y_S(t) = G_{SS}(t)$$

## A.8. Consequences of the Improved Modelling Approach

and system availability can be computed with equation A.11:

$$A_{SS}(t) = \frac{1}{10'000} \cdot \int_0^{10'000} Y_S(x)dx = 0.886$$

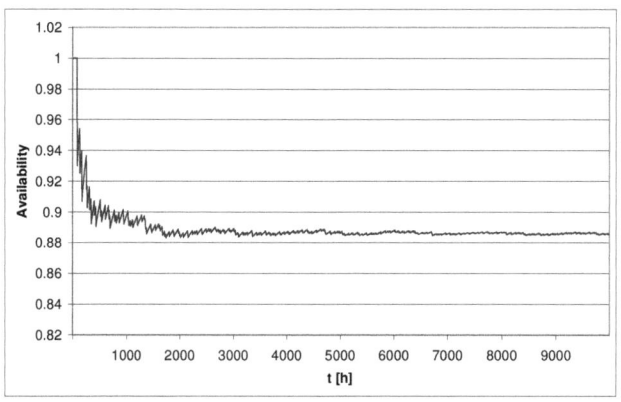

**Figure A.10:** *Cumulative System Availability*

In Figure A.10, system availability is calculated with $AnyLogic^©$ (a professional simulation tool for modelling complex hybrid, discrete and continuous systems) and plotted over time. After a sharp drop between $t = 0$ and $t = 1'000$ hours, system availability slightly increases and stabilizes at around 0.886. This stabilizing effect is not an indicator for reaching a steady-state of the system itself but simply derives from the computation of the system availability. Impact of single failures on system availability is mitigated with increasing time $t$ (see equation A.11). By contrast to system response, system availability stabilizes and reaches a steady state.

### A.8 CONSEQUENCES OF THE IMPROVED MODELLING APPROACH

The proposed methodology causes the following consequences:

**Simplicity of Calculation** System availability is represented in a simple, closed formula that can be computed with an ordinary computer and any math software.

**Efficient Algorithm** Proposed methodology overcomes the shortcomings of state space explosion caused by simultaneous events. Whereas simultaneous events provoke an exponential growth in combinatorial terms, complexity of the novel approach exhibits a linear increase. Calculation complexity is linearly dependent on the amount of incorporated components. Since the Laplace retransformation is performed on component level and not on system level as in multivariate renewal processes, solution can be found analytically and time-consuming numerical approximation techniques are avoided.

**Accurate System Availability Calculation** Classic approaches truncate simultaneous events; rare event approximation is applied most. This reduction of complexity causes a loss of precision in the system availability calculation resulting in an underestimate of system availability. Diminished accuracy may be of significance in large systems with components of low availability.

**Incorporating Buffers** Impact of buffers on system availability is absent in all other methods although their impact on system availability cannot be neglected. Suggested method offers an ease of use approach to model buffer effects on system characteristics.

**Dynamic System Availability** The novel method provides temporal estimations about expected system availability than classic static approaches by incorporating system dynamics. The quintessence is that system availability is no longer a simple fix number but is a dynamic measurement. Proposed methodology combines the ease-of-use of static methods with the higher accuracy of dynamic techniques.

The proposed methodology shall be regarded as an alternative to established availability methods as Block Diagramms with the benefit of increased accuracy and the possibility to represent temporal characteristics.

# Appendix B

# Queueing Theory

A queueing system consists of a buffer (queue) of specific capacity and one or more identical components (see Figure B). Queueing theory is a discipline of operations research and provides a collection of models to deal with the resulting effect of a buffer on a production system. They can only serve one job at a time and are either in a "busy" or "idle" state. If all components are "busy" when a new job arrives, this job is buffered and waits for its turn. According to a queueing discipline the next processing jobs is selected if one or more of the components are idle.

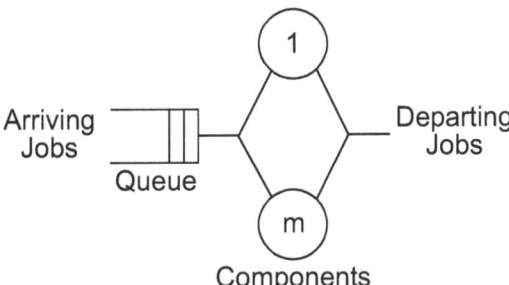

**Figure B.1:** *Queueing System*

The queueing system is characterized by its interarrival and service times, by its amount of components and by its buffer capacity. An encompassing introduction into the field of queueing theory can be found in [Bolch et al., 1998], [Ferschl, 1964] or [Siegle, 1995].

The measurand of interest is the expected waiting time of a job in the queue $E[WT]$:

$$E[WT] \simeq \frac{U(t,T_i)}{1-U(t,T_i)} \cdot \frac{1+CV^2[CT]}{2} \cdot \frac{CT}{m} \qquad \text{(B.1)}$$

$U(t,T_i)$ = Utilization of the production system during $T_i$
$CV$ = Ratio of standard deviation to mean of a distribution
$CT$ = Cycle time
$m$ = Amount of parallel components

As larger the expected waiting time of a job in a queue, as higher is the lead time and the work in process. However, high production system output and low lead time and work in process are conflictive objectives which can be proven for evidence by combining equation 3.3 with equation B.1. Equation B.1 can be solved for $U(t,T_i)$:

$$U(t,T_i) = \frac{E[WT] \cdot m}{E[WT] \cdot m + 0.5 \cdot CT \cdot (CV^2[CT]+1)} \qquad \text{(B.2)}$$

This can be inserted into equation 3.3:

$$O(t,T_i) = \frac{E[WT] \cdot m \cdot A_{Missions}(t,k,T_i) \cdot T_i}{CT \cdot E[WT] \cdot m + 0.5 \cdot CT^2 \cdot (CV^2[CT]+1)} \qquad \text{(B.3)}$$

High production system output entails large queues and therefore a lot of work in progress.

**High capacity utilization leads to large queues** Achieving high utilization and short queues/ lead times is a conflict of objectives. Queue length exponentially increases with higher capacity utilization.

**High capacity utilization leads to waiting time $\gg$ operation time** In case of high capacity utilization waiting time will be substantially larger than operation time.

**Lower variation of coefficient of $T_S$ leads to lower waiting time** Coefficient of variation has an over-proportional impact on the mean waiting time. Since the coefficient of variance is an indicator of uncertainty, low uncertainty in the mean service time brings about a low coefficient of variance and a low expected waiting time. Uncertainty in the mean service time is correlated to system reliability.

**Less operations leads to less queues** Fewer operations means fewer queues, respectively less interruptions of production processes leads to less buffers and queues.

**Decreased operation Time leads to less queues** A reduction of operation time decreases queue length. The same effect has an increase of the amount of redundant components.

Queueing theory provides some elementary insights into the way how queued production systems behave. However, it is not possible to directly apply quantitative results of queueing theory to practice, due to simplifications as independence between arrival and execution process, random arrival process or steady-state consideration. Queueing networks enhance the field of application for queueing theory to production systems with closed-loop designs but require Markov chains for their calculation. The deficiencies of Markov chains have been intensively discussed in section 2.6.

## Appendix C

# Demand Forecast

Forecasting and predicting are approaches to determine the most likely outcome of an uncertain variable in advance. For planning and controlling of logistics, a prediction about future demand is required due to the time lag in matching supply and demand [Lewis, 1997]. There exist no forecasting methods which are superior to others in every aspects. Every technique has its strengths and weaknesses. In order to calculate a reasonable forecast, demand must show some regularities or patterns (regular demand). This is the case when there are many small customers purchasing only a small fraction of the whole sales volume (single demand is asymptotically negligible) and $CV[P^{D(t,T_i)}] \leq 0.4$). Then, demand distribution follows a normal distribution and allows application of forecasting techniques. On the contrary, when demand is lumpy or irregular, there is so much uncertainty in the demand that a reliable forecast is difficult to make. Lumpy demand is often the case when some large customer dominate the demand pattern or when the volume of each item is low. There are two possible approaches to tackle lumpy demand:

- Increase stock of inventory
- Reduce lead time

However, it is assumed that demand is regular. Forecasting techniques can be classified in qualitative and quantitative methods.

### C.1 QUALITATIVE METHODS

Qualitative methods are based on experience or on market surveys and combine, sometimes, different forecasts by using simple mathematical tools. Lack of information or insufficient history, so that quantitative methods cannot be applied, are the main reasons to chose a qualitative technique. Thus, qualitative methods are often used to predict demand of new products or services.

Sales force assessment, market research and the Delphi method are the best known and widely applied qualitative methods and are briefly discussed (compare with [Ghiani et al., 2004]).

**Sales Force Assessment** Company salesman develop a forecast based on their experience and market knowledge

**Market Research** Interviews with potential consumers or users are the basis for demand forecast

**Delphi Method** Questionnaires are submitted to a panel of experts. Filled in questionnaires are evaluated and a new questionnaire is created based on the new findings and sent to the experts again. Delphi method ends when there is a common agreement between the experts.

If there is enough information available about demand history, quantitative methods can be used.

## C.2 Quantitative Methods

Those methods utilize demand history to identify possible regularities in the demand pattern by applying regression analysis or other mathematical techniques. Two categories of quantitative methods can be identified:

**Causal Methods** include regression, econometric models, input-output models, life-cycle analysis, computer simulation models and neural networks. However, most of these approaches are difficult to implement due to their complexity. This is the reason why only single or multiple regression is used for logistics planning and control in practice.

**Time Series Extrapolation** assumes that some regularities in the past will remain the same for the future. This pattern is then projected in the future. Some of those techniques are elementary technique, moving averages, exponential smoothing techniques or the decomposition approach.

## C.2. Quantitative Methods

Kind of historical data and the type of product determine the most suitable forecasting technique.
The following explanations are inspired by [Ghiani et al., 2004], [Schoensleben, 2002] and [Axsaeter, 2006].

### C.2.1 Causal Methods

If there exists a strong correlation between the future demand and the past values of some causal variables, this correlation can be used for demand forecasting. As an example, sales of economy cars is linked with the level of economic activity (Gross Domestic Product) or the demand for spare parts depends on the number of sold devices in the past using them. Causal methods can anticipate variations in demand but often the absence of adequate causal variables, that leads the forecast variable in time, makes the application of causal methods impossible.
Regression is a statistical method that associates a dependent variable $D(t, T_{i+1})$ (future demand) to a selection of causal variables $D(t, T_1), D(t, T_2), \ldots, D(t, T_i)$ with known values:

$$D(t, T_{i+1}) = f(D(t, T_1), D(t, T_2), \ldots, D(t, T_i))$$

This relation can be of any arbitrary form (linear or nonliner, quadratic, hyperbolic,...) and is selected according to their ability to interpolate the observations best.

### C.2.2 Time Series Extrapolation

Fundamental assumption behind time series extrapolation methods is that main features of past demand pattern will be reproduced in the future. Mostly, those methods are used for short- and medium-term predictions, where the probability of fundamental changes is low. The focus here is on elementary technique, moving averages and exponential smoothing techniques. Those techniques can be applied to three specific cases.

- Constant Trend Case
- Linear Trend Case
- Seasonal Effect Case

In the remainder of this chapter the constant trend case is assumed.

## C.2.3 Elementary Technique

The technique uses the classic tools of mathematical statistics, the mean of a sample $(M(t, T_i))$ and the standard deviation $\sigma(t, T_i)$. $P(t, T_i)$ is the demand forecast at time $t$ for period $T_i$ and $F(t, T_i)$ the actual demand in period $T_i$. It is assumed that the time horizon $T_{PHmax}$ has been divided into a finite number of equally long time periods. The length of a time period is $T_i$. The forecast for the first time period $i+1$ ahead is

$$P(t, T_{i+1}) = F(t, T_i)$$

This method creates a demand forecast with one period delay compared to the demand pattern.

## C.2.4 Moving Average Forecast

The moving average forecast technique considers the individual values of a time series as samples from parent population. This parent population is a sample distribution with *constant parameters*. Moving Average forecast method recalculates the moving average of the sample population according to the principle of the moving average. This technique is generally used to predict a more or less constant demand with no linear trends nor seasonality.

$$P(t, T_i) = M(t, T_i) = \frac{1}{t}\sum_{j=1}^{i} F(t, T_{i-j})$$

$$\sigma(t, T_i) = \sqrt{\frac{1}{i-1}\sum_{j=1}^{i}(F(t, T_{i-j}) - M(t, T_j))^2}$$

This method rates any demand in the past equally and reacts very reluctant to changes.

## C.2.5 First-Order Exponential Smoothing Forecast

If it would be wished to adapt the forecasting technique to actual demand, the demands for the last periods must be weighted more heavily, according to the principle of the weighted moving average. This ends up in a more reactive forecast technique. $G_{i-j}$

## C.2. Quantitative Methods

expresses the weighting of demand in the period $T_{i-j}$.

$$M(t, T_i) = P(t, T_i) = \frac{\sum G_{i-j} \cdot F(t, T_{i-j})}{\sum G_{i-j}}$$

With

$$G_{i-j} = \alpha \cdot (1-\alpha)^j$$
$$j = \text{age of the period}, 0 \leq j \leq \infty$$
$$G_{i-j} = \text{weight of the period demand with age j}$$
$$\alpha = \text{smoothing factor}, 0 < \alpha < 1$$
$$\sum_{i=1}^{\infty} G_y = \frac{\alpha}{1-(1-\alpha)} = 1$$

the equations for the mean value and the mean absolut deviation (MAD) can recursively derived [Schoensleben, 2002]:

$$M(t, T_i) = P(t, T_i) = \alpha \cdot F(t, T_i) + (1-\alpha) \cdot M(t, T_{i-1}) \quad \text{(C.1)}$$
$$MAD(t, T_i) = \alpha \cdot |F(t, T_i) - M(t, T_{i-1})| + (1-\alpha)^1 \cdot |F(t, T_{i-1}) - M(t, T_{i-2})| + \ldots$$
$$= \alpha \cdot |F(t, T_i) - M(t, T_{i-1})| + (1-\alpha) \cdot MAD(t, T_{i-1})$$

The choice of smoothing parameter $\alpha$ determines the weighting of current and past demands according to the given formulas. A high smoothing constant results in a rapid but also instantaneous reaction to changes in demand behavior. Therefore, the value for the smoothing parameter must be determined very carefully and it would be wishful to have a self-adapting smoothing factor.

### C.2.6 Trigg and Leach Adaptive Smoothing Technique

Adaptive smoothing is a form of exponential smoothing in which the smoothing constant is automatically adjusted as a function of forecast error measurement. If forecast values exceed the control limits, for example, $\pm |x \cdot \sigma|$ from the mean value, the model or the parameters must be altered. There's no consensus about the best adaptive approach. However, the most applied procedure was developed by Trigg and Leach in 1967 [Trigg and Leach, 1967]. They suggest the following method for continuous adjustment of the exponential smoothing parameter. The smoothing constant $\gamma$ smoothes forecast errors exponentially according to:

$$MD(t, T_i) = \gamma \cdot (F(t, T_i) - M(t, T_{i-1})) + (1-\gamma)^1 \cdot (F(t, T_{i-1}) - M(t, T_{i-2})) + \ldots$$
$$= \gamma \cdot (F(t, T_i) - P(t, T_{i-1})) + (1-\gamma) \cdot MD(t, T_{i-1})$$

Particularly when the mean changes, a large deviation results. In that case, a relatively large smoothing constant $\alpha$ should be chosen, so that the mean adjusts rapidly. In first-order exponential smoothing, it is reasonable to choose a smoothing constant that is related to the absolute value of the deviation. The result is a forecast formula with the variable smoothing constant $\alpha_t$ and a constant, relatively small $\gamma$ factor (in the range of $0.05 - 0.1$) to smooth forecast errors.

$$\alpha(t, T_i) = |AWS(t, T_i)| = \left| \frac{MD(t, T_i)}{MAD(t, T_i)} \right|$$

$$\alpha(t, T_i) = \left| \frac{\gamma \cdot (F(t, T_i) - P(t, T_{i-1})) + (1 - \gamma) \cdot MD(t, T_{i-1})}{\gamma \cdot |F(t, T_i) - M(t, T_{i-1})| + (1 - \gamma) \cdot MAD(t, T_{i-1})} \right|$$

This modified $\alpha(t, T_i)$ factor can be used in formula C.1 instead of the constant $\alpha$.

$$M(t, T_i) = P(t, T_i) = \alpha(t, T_i) \cdot F(t, T_i) + (1 - \alpha(t, T_i)) \cdot M(t, T_{i-1})$$

Unfortunately, this approach sometimes provides unstable forecasts. This deficiency was tried to overcome by limiting $\alpha(t, T_i)$ to a set of discrete values when certain control limits have been violated. For further discussion please see [Taylor, 2004].

## Appendix D
# Maintenance Strategy Optimization Procedures

Elaboration of a maintenance strategy optimization is a non-trivial issue since many different and partially contradictory requirements have to be incorporated. In this activity, decision aids can provide useful support facilitating, structuring and systematizing the optimization process.
Reliability-centered Maintenance (RCM) and Total Productive Maintenance (TPM) are two of a multitude of management approaches focused on establishing, servicing and optimizing a maintenance concept. They provide a framework to define a complete maintenance strategy. RCM and TPM are not novel maintenance strategies but they offer smart approaches to efficiently select the components to be maintained and arrange the different maintenance tasks. They support the finding and introducing of a maintenance strategy.

## D.1 RELIABILITY-CENTERED MAINTENANCE (RCM)

RCM is a method for planning maintenance activities developed in and for the airline industry which has been adapted to multiple other industries. Rausand [Rausand, 1998] defines RCM as:

**Definition 3** *Reliability-Centered Maintenance is a systematic consideration of system functions, the way functions can fail, and a priority-based consideration of safety and economics that iden-*

# Appendix D. Maintenance Strategy Optimization Procedures

*tifies applicable and effective preventive maintenance tasks.*

Resulting maintenance strategy is addressed to preserve the essential functions of the production system with cost-effective tasks. Back in the late 1960s, RCM was introduced in the airline industry. At this time, airlines were faced with a growing amount of maintenance requirements from Federal Aviation Administration (FAA), resulting in extensive maintenance plans with very expensive maintenance tasks. These maintenance plans were mainly addressed to the lately introduced Boeing 747 and would have caused costs that airlines would not have been able to operate the 747 profitable if they had to follow the FAA guidelines. Thus, a maintenance steering group was established to research a generally applicable approach for further developing maintenance strategies that could ensure maximum safety at reasonable costs. This research was the fundament for developments by Stanley Nowlan and Howard Heap which lead to the principles of RCM. Today, RCM is applied in many different industries such as nuclear power, chemical, automotive, manufacturing, etc. It is based on the knowledge that there are systems, which do not generally experience a wear-out phase and that not all failure concern the system equally (different criticality of sub-system breakdowns). Hence, a replacement or preventive maintenance strategy does not effect failure rate and will not prevent such type of failures. The idea behind RCM is not to achieve a maximum level of safety, reliability or availability of a system at any costs but to find a well-balanced combination of efficient and applicable maintenance tasks. Furthermore, emphasis in maintenance should be on monitoring and maintaining important system functions. These findings lead to two new criteria to decide if the maintenance task should be performed:

**Applicable preventive maintenance tasks** Applicable suggests that if this task is performed, it will prevent or mitigate a failure or could reveal a hidden failure.

**Effective preventive maintenance tasks** Effective means that the chosen maintenance task is the best (least expensive) task.

Basic assumption behind RCM is the believe that the inherent system's reliability is a function of the design and the built quality. Underlaying intension of RCM is to ensure that the inherent reliability is realized and not to improve the reliability of the system. Reliability can only be enhanced through redesign or modification, hence by changing the system. In particular RCM is not aimed to prevent any failure regardless of costs. It was designed to balance the costs and benefits to achieve a cost-effective preventive maintenance strategy and avoiding or removing unnecessary maintenance actions of

## D.1. Reliability-Centered Maintenance (RCM)

existing maintenance plans. For further discussions see [Moubray, 1991], [Nowlan and Heap, 1978] and [Smith, 1993]. A graphical depiction of the RCM process is represented in Figure D.1:

Process starts with defining the operating context of the production system, the definition of some minimal requirement standards and boundaries when this state is left (e.g. definition of the states "In Preventive Maintenance", "In Production" and "In Failure/Repair" in Figure 5.2). Based on this minimal standards, a Failure Mode Effects and Criticality Analysis (FMECA) is performed to identify the most critical components, their failure modes and the impact of their failure on the system. For components whose failures may have a severe influence on system level, adequate preventive maintenance tasks are elaborated which are aimed at avoid failure occurrence. Those PM tasks are rated according to their *applicability* and *effectiveness*. This rating is the essential part of the RCM approach. Only applicable and effective preventive maintenance tasks will be integrated in the maintenance strategy. If one of this two requirements is

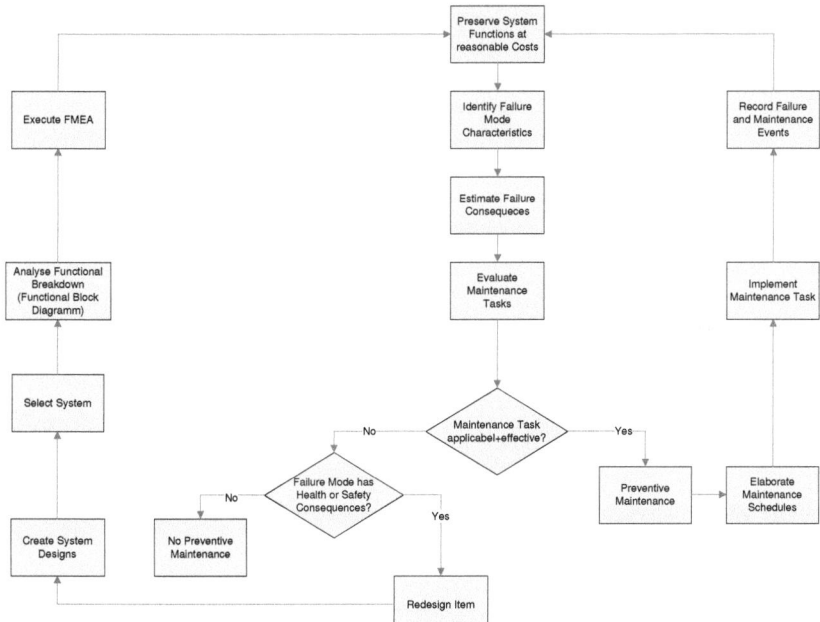

**Figure D.1:** *RCM-Process*

violated, it will be investigated if the failure occurrence may have any health or safety consequences. Components with health or safety risks need to be redesigned. The redesign process is depicted in the left branch in Figure D.1. All applicable and effective preventive maintenance tasks get scheduled and rationalized to be called up in work packages. Finally, maintenance strategy is periodically reevaluated and effectiveness tracked with help of the recorded maintenance and failure events.

## D.2 TOTAL PRODUCTIVE MAINTENANCE(TPM)

Whereas RCM is solely focused on the literal maintenance process, Total Productive Maintenance (TPM) takes a step forward by integrating managerial, organisational and production aspects into maintenance strategy optimization. TPM is a new way of understanding maintenance as integrative part of production system operating and improvement and is more a philosophy, a way of thinking, than a simple approach how to setup a maintenance strategy. Nakajima [Nakajima, 1988] defined TPM as following:

**Definition 4** *Total Productive Maintenance is a production system improvement methodology, which enables continuous and rapid improvement of the manufacturing process through the use of employee involvement, employee empowerment and closed-loop measurement of results. In a TPM program, teams comprised of operators, maintenance technicians, engineers and equipment suppliers are formed to improve the productivity of a key piece of equipment in the factory. The goals of a TPM program include:*

1. *Reduce manufacturing costs*
2. *Maximize the effective use of production system equipment (increase OEE, compare with equation D.2)*
3. *Increase the skills of the operations and maintenance personnel*
4. *Improve employee morale*

Developed and introduced by the Japan Institute of Plant Maintenance during 1970's, TPM is an offspring of the Total Quality Management (TQM) methodology. TQM is a

## D.2. Total Productive Maintenance(TPM)

management approach, centered on quality, using statistical analysis in manufacturing to control quality, and based on the participation of all employee to provide customer satisfaction. Some of the production system maintenance activities, performed as a part of the TQM program, couldn't be well integrated in the maintenance environment and often resulted in machines being "over-maintained" because of the thought "if a little oil helps, a lot should be better". During that time, there was little or no involvement of machine operator in setting up the maintenance tasks and schedules and maintenance program was created upon manufacturers recommendations. The need to improve productivity and quality led to modifications in the original TQM concepts, elevating maintenance of being an integral part of the overall quality program. This contribution from maintenance was labelled "Total Productive Maintenance". Since TPM is a descendent of the TQM concept, they share the same objectives: maximum customer satisfaction. On of the principle contributors to customer satisfaction is delivery on time which is strongly impacted by inventory and production system availability. If production system availability is not predictable, extra stock must be kept to buffer against this uncertainty. Thus, TPM is often regarded as a critical adjunct to lean manufacturing and JiT.

TPM incorporates more than only some methodologies and techniques but is considered being a philosophical approach how to perform and organize maintenance and linking production and maintenance department. The TPM concept includes the following steps:

- Maximize Overall Equipment Effectiveness (OEE) by eliminating the principle equipment losses (see equation D.1)

- Implement TPM in several departments as maintenance, operation, and production

- Involve every employee

- Establish small-group activities to motivate/educate employees

There has been a trend towards investigating the sources of insufficient production system output in the last few years, resulting in an new performance measurement called Overall Equipment Effectiveness (OEE). OEE was introduced to have a single value that incorporates the six major sources of "losses" - breakdowns, setup and adjustment, small stops, reduced speed, startup rejects and production rejects (see Figure D.2). These six losses refer to three general categories - down time loss $= \left(\frac{B}{A}\right)$, speed loss $= \left(\frac{D}{C}\right)$ and quality loss $= \left(\frac{F}{E}\right)$ (see equation D.1 and [de Ron and Rooda, 2005], [Kwon and Lee, 2004]).

# Appendix D. Maintenance Strategy Optimization Procedures

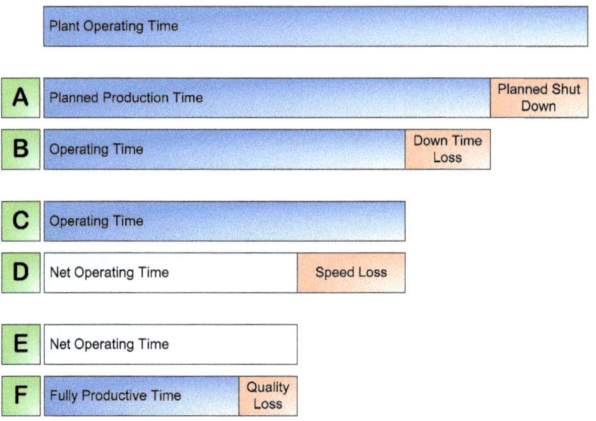

**Figure D.2:** *Overall Equipment Effectiveness*

$$OEE = \frac{B}{A} \cdot \frac{D}{C} \cdot \frac{F}{E} \tag{D.1}$$

Destroying barriers between the maintenance and the production department is considered as main contribution of TPM to increase the company's efficiency. Operators field of activities is enlarged from only operating equipment to monitoring the condition of the equipment and breakdown prevention. Production and maintenance crew are jointly responsible for preventing equipment from failure. Whereas maintenance department is released from performing simple maintenance tasks, production crew broadens its activities, resulting in an improved equipment effectiveness (shortened time for reaction, lower labour costs, etc.). TPM is extensively discussed in literature, for further details please [Nakajima, 1988], [Willmott and McCarthy, 2001], [Al-Radhi, 2002] or [Suzuki, 1994].

VDM Verlagsservicegesellschaft mbH

Die VDM Verlagsservicegesellschaft sucht für wissenschaftliche Verlage abgeschlossene und herausragende

## Dissertationen, Habilitationen, Diplomarbeiten, Master Theses, Magisterarbeiten usw.

für die kostenlose Publikation als Fachbuch.

Sie verfügen über eine Arbeit, die hohen inhaltlichen und formalen Ansprüchen genügt, und haben Interesse an einer honorarvergüteten Publikation?

Dann senden Sie bitte erste Informationen über sich und Ihre Arbeit per Email an *info@vdm-vsg.de*.

**Sie erhalten kurzfristig unser Feedback!**

VDM Verlagsservicegesellschaft mbH
Dudweiler Landstr. 99    Telefon  +49 681 3720 174
D - 66123 Saarbrücken    Fax      +49 681 3720 1749
**www.vdm-vsg.de**

Die VDM Verlagsservicegesellschaft mbH vertritt

Printed by Books on Demand GmbH, Norderstedt / Germany